Occurrence and Environmental Risks of Organic Pollutants in Aquatic Environment

Occurrence and Environmental Risks of Organic Pollutants in Aquatic Environment

Guest Editors

Chenglian Feng
Bingli Lei
Weiying Feng

MDPI

Basel • Beijing • Wuhan • Barcelona • Belgrade • Novi Sad • Cluj • Manchester

Guest Editors

Chenglian Feng
State Key Laboratory of Environmental Criteria and Risk Assessment
Chinese Research Academy of Environmental Sciences
Beijing
China

Bingli Lei
College of Environmental and Chemical Engineering
Shanghai University
Shanghai
China

Weiying Feng
School of Materials Science and Engineering
Beihang University
Beijing
China

Editorial Office
MDPI AG
Grosspeteranlage 5
4052 Basel, Switzerland

This is a reprint of the Special Issue, published open access by the journal *Toxics* (ISSN 2305-6304), freely accessible at: https://www.mdpi.com/journal/toxics/special_issues/86BS407D91.

For citation purposes, cite each article independently as indicated on the article page online and as indicated below:

Lastname, A.A.; Lastname, B.B. Article Title. *Journal Name* **Year**, *Volume Number*, Page Range.

ISBN 978-3-7258-7563-4 (Hbk)
ISBN 978-3-7258-7564-1 (PDF)
https://doi.org/10.3390/books978-3-7258-7564-1

Cover image courtesy of Weiying Feng

Contents

About the Editors

Chenglian Feng

Chenglian Feng is a researcher at the State Key Laboratory of Environmental Criteria and Risk Assessment, part of the Chinese Research Academy of Environmental Sciences in Beijing, China. Her research is deeply rooted in understanding the interactions between chemical contaminants and aquatic ecosystems, with a primary focus on ecotoxicology, environmental chemistry, and risk assessment. Dr. Feng's work is pivotal in developing the scientific underpinnings for environmental quality standards, aiming to translate complex toxicological data into practical criteria for protecting aquatic life. Her contributions to this Special Issue, particularly her collaboration on studies investigating combined pollutant toxicity and the derivation of tissue-residue guidelines for persistent organic pollutants like dieldrin, exemplify her commitment to advancing methodologies for more precise and ecologically relevant risk evaluations.

Bingli Lei

Bingli Lei is an Associate Professor at the Institute of Environmental Pollution and Health within the College of Environmental and Chemical Engineering at Shanghai University, China. Her research is dedicated to the identification, behavior, and toxicological effects of emerging contaminants in water environments. Dr. Lei's specific interests center on pharmaceuticals and personal care products (PPCPs) and emerging environmental endocrine disruptors, with a particular emphasis on bisphenol compounds. She is actively involved in screening for these typical endocrine disruptors and elucidating their potential impacts on aquatic organisms and, by extension, human health. Her work often combines field monitoring with in vitro bioassays to assess the risks posed by these complex chemical mixtures, contributing vital knowledge on their occurrence and mechanisms of action, as highlighted by her collaborative studies within this Special Issue.

Weiying Feng

Weiying Feng is an Associate Professor in the School of Materials Science and Engineering at Beihang University in Beijing, China. Her research addresses critical challenges in water pollution and environmental control, with a strong foundation in environmental ecotoxicology. Dr. Feng's work employs interdisciplinary approaches to investigate the fate and transport of pollutants in aquatic systems. Her contributions to the field, including studies on phosphorus fractions in lake sediments and the ecological shifts that influence nutrient release in plateau lakes like Dianchi and Erhai, provide essential insights for managing eutrophication and understanding complex pollutant dynamics. Through her research, she aims to develop strategies for pollution control that are informed by a deep understanding of both chemical and ecological processes in impacted water bodies.

toxics

MDPI

Editorial

Environmental Behaviors, Ecological Risks, and Toxic Mechanisms of Emerging and Legacy Contaminants in China: From Distribution to Management

Weiying Feng [1], Chenglian Feng [2,*] and Bingli Lei [3]

[1] School of Materials Science and Engineering, Beihang University, Beijing 100191, China; fengweiying@buaa.edu.cn
[2] State Key Laboratory of Environmental Criteria and Risk Assessment, Chinese Research Academy of Environmental Sciences, Beijing 100012, China
[3] Institute of Environmental Pollution and Health, College of Environmental and Chemical Engineering, Shanghai University, Shanghai 200444, China; leibingli@126.com
* Correspondence: fengcl@craes.org.cn

1. Introduction

In recent years, the accelerated pace of industrialization and urbanization has exacerbated the coexistence and combined pollution of both emerging and legacy contaminants in aquatic environments, posing potential threats to ecosystem integrity and human health. This Special Issue, entitled "Occurrence and Environmental Risks of Organic Pollutants in Aquatic Environment", compiles eight research articles focusing on the environmental behaviors, toxic effects, and risk assessment of various contaminants, including polycyclic aromatic hydrocarbons (PAHs) (contribution 1), per- and polyfluoroalkyl substances (PFASs) (contribution 2), organophosphate flame retardants (OPFRs) (contribution 2), phosphorus fractions in lake sediments (contribution 3), microplastics (contribution 4), disinfection byproducts (DBPs) (contribution 5), the plasticizer DEHP (contribution 6), pesticides (contribution 7), and the persistent organic pollutant dieldrin (contribution 8).

Through the application of field monitoring, bioassays, bioinformatics, metabolomics, and species sensitivity distribution models, these studies systematically elucidate the multimedia distribution, bioaccumulation potential, and toxicological effects of these pollutants, including neurotoxicity, immunotoxicity, oxidative stress, and ecological risks in aquatic organisms. The findings provide a robust scientific basis for the integrated management of both emerging and traditional pollutants in aquatic environments. Moreover, they directly contribute to the refinement of global water quality criteria and the prioritization and precise identification of high-risk contaminants while offering key technical guidance for lake eutrophication control, drinking water source protection, and watershed-scale ecological risk management.

2. Overview of Published Articles

To begin with, Li et al. systematically investigated the distribution and ecological risks of polycyclic aromatic hydrocarbons (PAHs) and pesticides in the surface water and sediments of the Danjiangkou Reservoir. The results showed that the total concentration of PAHs (ΣPAHs) in the water column ranged from 64.64 to 868.23 ng/L, with low-molecular-weight PAHs (e.g., naphthalene, fluorene, and phenanthrene) being the predominant compounds. Pesticides were mainly composed of organophosphorus compounds, with ΣOPPs concentrations ranging from 2.62 to 71.14 ng/L. In sediments, ΣPAHs concentrations ranged from 0.01 to 2.93 ng/g, dominated by high-molecular-weight PAHs, while organophosphorus pesticides such as malathion and chlorpyrifos exhibited relatively higher concentrations. Ecological risk assessment indicated that phenanthrene, malathion, and other contaminants posed moderate to high risks to aquatic organisms, particularly to daphnids, whereas human health risks were relatively low. The study concludes that although the Danjiangkou Reservoir is generally safe as a drinking water source, certain organic pollutants present potential threats to the aquatic ecosystem, underscoring the need for continuous monitoring (contribution 1).

Building on this, Sun et al. further characterized the occurrence of 18 per- and polyfluoroalkyl substances (PFASs) and 9 organophosphate flame retardants (OPFRs) in the upper reaches of the Yangtze River. The results revealed that total PFAS concentrations ranged from 16.07 to 1231.11 ng/L, with ultra-short-chain PFASs (e.g., TFMS) being the most abundant. Total OPFR concentrations ranged from 17.36 to 190.42 ng/L, with halogenated OPFRs such as TCPP dominating. Pollutant levels were higher in the main stream than in tributaries, and contamination loads were greater during the rainy season. Ecological risk assessment indicated low risk at most sites, though moderate risks from OPFRs were identified in certain areas (e.g., Leshan and Chongqing), with wastewater discharge being a key source. Health risk assessment showed acceptable non-carcinogenic risks, but potential carcinogenic risks from OPFRs were identified for adults in the Leshan section. The study recommends strengthening regulatory oversight of emerging contaminants and improving wastewater treatment technologies (contribution 2).

In addition to organic pollutants, the transport and transformation of nutrients remain a central focus in aquatic environmental research. Wu et al. investigated the relationship between phosphorus fractions in sediments and dissolved organic matter (DOM), human activities, and ecological characteristics in three plateau lakes of Yunnan Province: Dianchi Lake (eutrophic, algae-dominated), Erhai Lake (mesotrophic, grass–algae mixed), and Yangzonghai Lake (mesotrophic, algae-dominated). The results demonstrated that human activities significantly promoted the accumulation of immobile phosphorus fractions (e.g., NaOH-rP and HCl-P) in sediments, while showing no significant effect on mobile phosphorus. Mobile phosphorus content was highest in Dianchi Lake, exceeding that in Erhai and Yangzonghai Lakes by 40.6% and 278.3%, respectively, indicating that the shift from macrophyte-dominated to algae-dominated states enhances phosphorus mobility. DOM influenced phosphorus mobility primarily by promoting the adsorption of iron-bound phosphorus (BD-P) and organic phosphorus (NaOH-nrP). The study emphasizes the need for differentiated management strategies tailored to lakes of varying trophic states to control internal phosphorus release (contribution 3).

Building on the understanding of pollutant distribution and transport, elucidating their degradation processes is essential for assessing environmental fate. Cao et al. employed bibliometric analysis to systematically review research progress on the biodegradation of microplastics from 2012 to 2022. The findings indicate that research hotspots have centered on functional microorganisms—such as bacteria from the phyla Proteobacteria and Firmicutes, and fungi from Ascomycota—alongside the mechanisms of action of their secreted degradative enzymes, including hydrolases and oxidases. Microorganisms degrade microplastics through colonization, biofilm formation, and enzymatic breakdown. Future research should expand the focus to include degradation mechanisms in soil and atmospheric environments and further explore the molecular degradation pathways of these functional microorganisms (contribution 4).

In addition to degradation processes, the direct toxic effects of pollutants and their underlying molecular mechanisms warrant in-depth investigation. Zhang et al. elucidated the toxic mechanisms of the disinfection byproduct 2,6-dichloro-1,4-benzoquinone (2,6-DCBQ)

https://doi.org/10.3390/toxics14030263

on *Microcystis aeruginosa*. The study revealed that 2,6-DCBQ inhibits algal growth, reduces photosynthetic pigment and protein contents, induces oxidative stress (elevated reactive oxygen species levels and altered antioxidant enzyme activities), and disrupts cell membrane integrity. Concurrently, it stimulated the synthesis and release of microcystin-LR (MC-LR) and regulated the expression of related genes (mcyA, mcyD, mcyH). Non-targeted metabolomics identified 208 differentially abundant metabolites, which were primarily enriched in pathways associated with ABC transporters, two-component systems, and folate biosynthesis. The study concludes that 2,6-DCBQ exerts comprehensive toxicity on *M. aeruginosa* through interference with energy metabolism, induction of oxidative damage, and promotion of toxin release, providing a theoretical basis for assessing the aquatic ecological risks of such emerging disinfection byproducts (contribution 5).

Similarly, the toxic mechanisms of plasticizers have garnered considerable attention. Meng et al. investigated the neurotoxicity and potential carcinogenic mechanisms of the plasticizer di-2-ethylhexyl phthalate (DEHP) using bioinformatics analyses and zebrafish toxicology assays. The results demonstrated that the circadian regulator PER3 was significantly downregulated in glioblastoma multiforme (GBM) tissues and was closely associated with immune cell infiltration, immune checkpoint genes, and oncogenes. Low PER3 expression was significantly correlated with poor prognosis and advanced clinical stage, exhibiting high diagnostic value (AUC = 0.916). Further zebrafish experiments revealed that DEHP exposure led to developmental abnormalities, increased anxiety-like behaviors, and significant downregulation of PER3 gene expression in brain tissue. The study suggests that PER3 is a potential novel diagnostic marker for GBM, and that DEHP may exert carcinogenic effects by disrupting circadian gene expression (contribution 6).

In real-world environments, pollutants often co-exist as complex mixtures, rendering their combined toxic effects considerably more intricate. Liu et al. investigated the acute toxicity, neurotoxicity, immunotoxicity, and behavioral effects of deltamethrin (DM) and sulfamethoxazole (SMX) on adult zebrafish. The results showed that the 96 h LC50 of DM was 4.84 μg/L, indicating high acute toxicity; however, following combined exposure with SMX, the LC50 increased to 11.32 μg/L, suggesting that SMX may partially mitigate the toxicity of DM. Exposure to DM significantly reduced the levels of GABA, 5-HT, and AChE in the brain, induced oxidative stress (elevated SOD and GST activities, decreased CAT and GPX activities), and suppressed the expression of immunoglobulin genes (IgM, IgD, IgZ), while upregulating the pro-inflammatory cytokine IL-8. Combined exposure partially restored AChE activity and feeding behavior. This study reveals that DM exhibits potent neurotoxicity and immunotoxicity, and that the co-presence of SMX can modulate these toxic effects, highlighting the need to consider chemical interactions in environmental risk assessments (contribution 7).

To translate toxic effects into actionable management thresholds, the derivation of scientifically sound water quality criteria is essential. Xie et al. employed a tissue residue approach, combined with a bioaccumulation factor (BAF) model, to derive water quality criteria for dieldrin aimed at protecting aquatic organisms and wildlife. By screening tissue-based toxicity data for aquatic organisms, birds, and mammals, tissue reference values were obtained using the species sensitivity distribution (SSD) method. These values were then back-calculated to corresponding water concentrations using an estimated BAF of 58,884.37 L/kg. The results indicated that the criteria for protecting aquatic life and aquatic-dependent wildlife are 3.86 ng/L and 1.4 ng/L, respectively. Risk assessment revealed that dieldrin residues in some surface waters of China pose potential ecological risks to aquatic organisms and birds. The study recommends adopting the tissue residue approach to derive water quality criteria for bioaccumulative pollutants, thereby ensuring more effective protection of aquatic ecosystems (contribution 8).

https://doi.org/10.3390/toxics14030263

In summary, these studies collectively reveal the complex reality of aquatic environments, characterized by a diverse array of pollutants, multiple sources, and varied toxicological mechanisms. For instance, research on the Danjiangkou Reservoir demonstrated that, despite relatively low overall concentrations of PAHs and pesticides, certain organophosphorus pesticides pose moderate to high risks to aquatic organisms (contribution 1). Investigations in the upper reaches of the Yangtze River revealed the widespread occurrence of PFASs and OPFRs, with wastewater discharge serving as a significant source and potential carcinogenic risks identified at specific sites (contribution 2). Studies on plateau lakes such as Dianchi and Erhai further indicated that human activities substantially influence phosphorus speciation and mobility in sediments, and that the ecological shift from macrophyte-dominated to algae-dominated states exacerbates the risk of endogenous phosphorus release (contribution 3). Furthermore, research on the combined toxicity of deltamethrin and sulfamethoxazole highlighted that interactions between co-occurring pollutants can alter the toxic effects of individual compounds, underscoring the necessity of considering mixture effects in risk assessments (contribution 7). More importantly, the derivation of water quality criteria for dieldrin using a tissue residue approach (contribution 8) addresses the limitations of traditional water-based criteria in adequately protecting against bioaccumulative pollutants, thereby offering a novel strategy for the precise control of high-risk persistent contaminants.

3. Future Perspectives and Concluding Remarks

Collectively, the research presented in this Special Issue has not only deepened the understanding of the fate and toxic mechanisms of both legacy and emerging contaminants in aquatic environments but has also provided a critical scientific basis for refining China's water quality criteria system, optimizing pollutant monitoring and risk assessment methodologies, and formulating differentiated watershed-scale pollution control strategies. These studies have collectively advanced a paradigm shift from single-pollutant assessment to combined pollution evaluation, from external exposure concentrations to internal exposure doses, and from water-based criteria to tissue-residue-based criteria, thereby laying a solid foundation for the precise protection of aquatic ecosystems and the achievement of sustainable development goals.

Future research should be directed toward the following key areas: first, further investigation into the combined toxic mechanisms of multiple pollutants under co-exposure scenarios, with a focus on synergistic or antagonistic interactions and their integrated effects on the nervous, immune, and metabolic systems of aquatic organisms. Second, deeper exploration of molecular mechanisms underlying toxicity, integrating multi-omics technologies to elucidate toxicological pathways and identify robust biomarkers. Third, the refinement of water quality criteria derivation methods based on tissue residue and bioaccumulation assessments to enhance the precise control of persistent and bioaccumulative pollutants. Fourth, the expansion of research matrices from the aqueous phase to sediments, soils, and the atmosphere, thereby build a multimedia integrated ecological risk assessment framework. Ultimately, these efforts will establish full-chain scientific support—from pollutant identification to toxic mechanism elucidation and management application—providing both theoretical foundations and technical guidance for aquatic ecosystem health and sustainable watershed management.

Author Contributions: W.F.: writing—original draft preparation; C.F. and B.L.: writing—review and editing. All authors have read and agreed to the published version of the manuscript.

https://doi.org/10.3390/toxics14030263

Funding: This research was jointly supported in part by the Key Research and Development and Technology Transfer Program of Inner Mongolia (2025YFHH0145) and the Science and Technology Major Project of Ordos City (ZD20232301).

Data Availability Statement: Data sharing is not applicable.

Acknowledgments: The Guest Editor would like to express sincere gratitude to all the authors for their valuable contributions to this Special Issue, as well as to the reviewers for their time and expertise in evaluating the manuscripts. The editorial team of *Toxics* are also acknowledged for the administrative and technical support they provided throughout the process.

Conflicts of Interest: The authors declare no conflicts of interest.

List of Contributions

1. Li, R.; Pang, H.; Guo, Y.; Zhou, X.; Fu, K.; Zhang, T.; Han, J.; Yang, L.; Zhou, B.; Zhou, S. Distribution of Polycyclic Aromatic Hydrocarbons and Pesticides in Danjiangkou Reservoir and Evaluation of Ecological Risk. *Toxics* **2024**, *12*, 859. https://doi.org/10.3390/toxics12120859.
2. Sun, W.; Fu, Z.; Liu, Y.; Bai, Y.; Zhao, Y.; Wang, C.; Wu, F. Per- and Poly-Fluoroalkyl Substances, and Organophosphate Flame Retardants in the Upper Yangtze River: Occurrence, Spatiotemporal Distribution, and Risk Assessment. *Toxics* **2025**, *13*, 116. https://doi.org/10.3390/toxics13020116.
3. Wu, X.; Wang, Y.; Jiao, L.; He, J.; Zhou, H.; Hao, Z. Influencing Factors of Phosphorus Mobility and Retention in the Sediment of Three Typical Plateau Lakes. *Toxics* **2025**, *13*, 120. https://doi.org/10.3390/toxics13020120.
4. Cao, Y.; Bian, J.; Han, Y.; Liu, J.; Ma, Y.; Feng, W.; Deng, Y.; Yu, Y. Progress and Prospects of Microplastic Biodegradation Processes and Mechanisms: A Bibliometric Analysis. *Toxics* **2024**, *12*, 463. https://doi.org/10.3390/toxics12070463.
5. Zhang, T.; Wang, Z.; Wu, L.; Liu, C.; Meng, L.; Tian, F.; Hou, M.; Lin, H.; Ye, J. The Effects of Disinfection Byproduct 2,6-Dichloro-1,4- benzoquinone on the Cyanobacterium *Microcystis aeruginosa*: From the Perspectives of Biochemistry and Non-Targeted Metabolomics. *Toxics* **2025**, *13*, 64. https://doi.org/10.3390/toxics13010064.
6. Men, S.; Xu, J.; Yang, Z.; Yan, Z. DEHP-Induced Glioblastoma in Zebrafish Is Associated with Circadian Dysregulation of PER3. *Toxics* **2024**, *12*, 835. https://doi.org/10.3390/toxics12120835.
7. Liu, Y.; Liu, F.; Wang, C. Acute Toxicity, Neurotoxic, Immunotoxic, and Behavioral Effects of Deltamethrin and Sulfamethoxazole in Adult Zebrafish: Insights into Chemical Interactions and Environmental Implications. *Toxics* **2025**, *13*, 128. https://doi.org/10.3390/toxics13020128.
8. Xie, L.; Li, X.; Bao, L.; Zhang, Y.; Su, H.; Liu, X.; Wang, F.; Wei, Y.; Ji, N.; Zhou, M. Water Quality Criteria of Dieldrin for the Protection of Aquatic Organisms and Wildlife Using a Tissue Residue Approach. *Toxics* **2025**, *13*, 173. https://doi.org/10.3390/toxics13030173.

toxics

MDPI

Article

Water Quality Criteria of Dieldrin for the Protection of Aquatic Organisms and Wildlife Using a Tissue Residue Approach

Li Xie [1], Xuemei Li [1], Liangwen Bao [1,2], Yuexin Zhang [1], Hailei Su [1,*], Xuesong Liu [1], Fanfan Wang [1], Yuan Wei [1], Ningning Ji [1] and Min Zhou [1]

1 State Key Laboratory of Environmental Criteria and Risk Assessment, Chinese Research Academy of Environmental Sciences, Beijing 100012, China
2 College of Environmental Science and Engineering, Liaoning Technical University, Fuxin 123032, China
* Correspondence: suhailei666@163.com; Tel.: +86-15001296136

Abstract: Dieldrin is legacy organochlorine insecticide, which was listed in the Stockholm Convention because of its persistence, bioaccumulation and toxicity. However, it is still present in the environment and in organisms two decades after its ban. The current criteria used for risk assessment in China are based on acute toxicity data in water columns without considering the bioavailability and bioaccumulation, which accordingly lead to the under-protection of aquatic organisms and wildlife. In this study, the water quality criteria (WQC) for dieldrin were derived from a combination of tissue-based toxicity data and the bioaccumulation factor (BAF) to better protect aquatic ecosystems. The dieldrin residue data in surface water in China were obtained by literature review and the ecological risk was assessed using the quotient method. Combined with a BAF of 58,884.37 L/kg estimated by the model, the WQC were calculated as needing to be 3.86 and 1.4 ng/L to protect aquatic life and aquatic-dependent wildlife, respectively. The results of the risk assessment revealed the potential high risk posed by dieldrin bioaccumulation. This study provides scientific guidance for the determination of the water quality standard for dieldrin and to ensure the risk management of the aquatic environment in China.

Keywords: dieldrin; water quality criteria; bioaccumulation; ecological risk; wildlife criteria

1. Introduction

Dieldrin (C12H8Cl6O) is both a synthetic organochlorine pesticide (OCP) classified as a chlorinated cyclodiene compound that was primarily used on cotton, corn, and citrus crops and an environmental and metabolic product of the related compound aldrin [1]. Dieldrin has a low water solubility of 0.110 mg/L at 20 °C [2]. It is lipophilic and persistent in soil has a half-life of up to five years. Dieldrin possesses high potential for bioaccumulation and readily bioaccumulates in terrestrial and aquatic organisms [3–6], tending to bio-magnify in the food chain.

The endpoints of dieldrin have been identified, including its carcinogenicity, endocrine disruptions, reproductive/development effects and neurotoxicity [7–10]. Due to its adverse effects on humans and the ecosystem, dieldrin's persistence in the environment has become a major concern, prompting its inclusion as one of the 12 initial persistent organic pollutants (POPs) under the Stockholm Convention, which imposed a ban on its manufacturing, usage and commerce worldwide. However, dieldrin has been detected worldwide in the environment through studies related to pesticides [11–15]. Its residue has also been monitored in human milk [16–18] and wildlife, such as in raptors across the world [19] and in whales and polar bears in the Arctic [20].

 https://doi.org/10.3390/toxics13030173

Dieldrin is very toxic to aquatic life with long lasting effects and is suspected of causing cancer. Some guidelines and standards have been set for dieldrin. In Canada, water quality guideline of the long-term concentration for the protection of aquatic life was first published in 1987 but is no longer recommended because environmental exposure predominantly occurs via sediment, soil and/or tissue. Only sediment quality guidelines for the protection of aquatic life are recommended. CCME developed interim sediment quality guidelines for Canada regarding the probable effect levels for dieldrin, which can be used to assess the likelihood that exposure to dieldrin in sediments will have detrimental biological effects [21]. The USEPA released the procedures to obtain dieldrin equilibrium partitioning sediment benchmarks, in order to protect benthic organisms [22]. The USEPA does not provide the ambient water quality criterion for dieldrin. But the wildlife value for dieldrin in waters within the Great Lakes Basin was calculated as 0.071 ng/L, based on field measurements and trophic level-specific factors [23]. In the European Union, the environmental quality standard (EQS) value for dieldrin, namely 10 ng/L, is used to evaluate the pollution of rivers [24]. Moreover, the EQS value has been amended to 10 ng/L for cyclodiene pesticides (including aldrin, dieldrin endrin, and isodrin) because these substances are not priority substances. In addition, the Atlantic RBCA states an EQS value of 1 ng/L of dieldrin for surface freshwater.

Dieldrin has also been detected in the Chinese environment, such as in soil, air, and water, as well as in wild animals [25–29]. In China, dieldrin was never produced on an industrial scale and was not used as a pesticide in agriculture [30]. However, previous studies have unequivocally demonstrated its presence, indicating the long-range transport of dieldrin. Recent research which was conducted in two sediment cores from Lake Fuxain, the largest deep freshwater lake in China, has found that dieldrin might pose a potential threat to exposed organisms [31]. Dieldrin is not included in China's environmental quality standards for surface water (GB3838-2002) [32]. A maximum concentration of dieldrin, 470 ng/L, was identified using acute toxicity data for domestic aquatic organisms in China [33]. However, dieldrin in aquatic environments generally occurs at low concentrations, and it tends to be found at high levels within living organisms, which is harmful and toxic. Therefore, it is necessary to identify the long-term concentrations of dieldrin for the protection of aquatic life. Moreover, the established water quality standards in China do not consider the trophic magnification of chemical substances and their effects at higher trophic levels. Data on wildlife are urgently needed to assess the risks posed by persistent, bioaccumulative and toxic chemicals (PBTs) to birds or mammals in Chinese aquatic systems and to support national policy-making decisions.

Traditional water quality criteria (WQC) derived from water-based toxicity that do not consider bioaccumulation factors (BAFs) make it challenging to provide comprehensive protection for aquatic life. In addition, multiple routes of exposure (e.g., diet, sediment, water) may impact the bioavailability of these substances. Accordingly, adopting a tissue-based method for deriving aquatic life criteria is recommended [34]. There are some advantages to using tissue concentrations or residues as the dose metric to characterize toxicity: species differences in bioaccumulation and time to effect are less important factors and exposure over time and space is integrated [35]. This method has been used to derive criteria for mercury, selenium, DDT and TCDD [36–39].

The aim of this study is to determine the WQC for dieldrin in order to protect aquatic organisms and aquatic-dependent wildlife using tissue-based toxicity data to better manage the environment. This study gives new insight into the ecological risk of dieldrin and provides a scientific foundation for enhancing WQC determination systems in China by serving as a reference for the derivation of bio-accumulative substances.

2. Methods

2.1. Toxicity Data Screening

The sources of tissue-based toxicity of aquatic organisms for dieldrin are the Environmental Residue Effects Database, the ECO-TOX database and the published literature. For wildlife, tissue-based toxicity data were obtained from the literature compiled by the USEPA. The regions of different species were mainly defined according to the Global Biodiversity Information Facility (https://www.gbif.org/, accessed on 18 September 2024). A tissue-based approach needs to be developed based on ecologically relevant endpoints that protect populations [35]. Thus, data selection is guided by the following principles: (1) internationally recognized, indigenous and introduced species, including freshwater invertebrates, vertebrates and wildlife were selected; (2) the endpoints of the effect include morality, developmental growth, reproduction, behavior and accumulation; (3) the no-observed effect concentration (NOEC) was selected as the test endpoint to develop chronic criteria; and (4) residue data are obtained from whole-body measurements.

2.2. Derivation of Tissue-Based Criteria

Provided that sufficient valid toxicity data are available, the species sensitivity distribution (SSD) method was selected to obtain a baseline or toxicity reference value. This probability distribution function assumes that sensitive species can be characterized by statistical distributions. The SSD method is widely used to obtain water quality benchmarks, characterizing the impacts of chemical contaminants on water quality and/or for ecological risk assessment [40]. Designating the most sensitive species, a hazard concentration, i.e., the 5th percentile of the SSD which protects 95% of organisms, is used to set the criteria [41].

The tissue-based criteria (TBC) refers to criteria generated from toxicity data reported as concentrations in tissues of the target organisms or their diet (in wildlife) [42]. Tissue-based toxicity metrics are developed using the same methodology as that for external concentration-based toxicity tests [35]. In the case of aquatic-dependent wildlife, toxicity data of avian or mammalian species expressed as tissue concentrations per day are used to generate HC_5, representing the tolerable daily intake (TDI). In order to obtain the WQC, tissue concentrations need to be represented as concentrations in the aquatic diet. The tissue reference value (TRV) for aquatic-dependent wildlife is calculated using the TDI in conjunction with the daily food ingestion rates (FIs) and the body weights (BWs) of wildlife species [43], as shown in Equation (1).

$$\mathrm{TRV} = \frac{\mathrm{TDI}}{\mathrm{FI} : \mathrm{BW}} \tag{1}$$

where the TDI is derived using the SSD method using tissue-based toxicity data (mg/kg) and FI:BW is the ratio of food ingestion rate to body weight of the representative wildlife species in China, which were selected from the literature.

2.3. Obtaining the WQC for Dieldrin

In order to translate the tissue criterion to corresponding water concentrations, it should be divided by the relevant bioaccumulation factors (BAFs) determined for each representative species [42]. Bioaccumulation can be viewed as the combination of bioconcentration and food uptake as it occurs when chemicals accumulate in aquatic organisms through multiple exposure routes, including dietary, respiratory, and dermal absorption [44]. The BAF is the ratio of the concentration of a chemical in an organism to that in the environment. Fish have among the highest bioaccumulation rates for dieldrin and higher trophic levels in aquatic ecosystems. From there, the empirical model based on field-derived BAFs is used to assess bioaccumulation potential and to translate tissue-based criteria to water concentrations [34]. There are no available site-specific BAF measurements obtained though paired water concentrations and fish tissue data for the recommended representative species. Herein, a generic BAF model for fish development by Arnot and Gobas [45] was used to estimate the BAF of dieldrin. The model provides predictions of BAF for fish species in the lower, middle and upper trophic levels of aquatic food webs and can be used to predict dietary concentrations for higher trophic level predators (e.g., birds and mammals) via the fish in their diet. The model is a quantitative structure–activity relationship (QSAR), with input parameters consisting only of the octanol–water partition coefficient (Kow) of the chemical and, if available, the metabolic transformation rate constant. Since the metabolic transformation rate constant was not available, the $\log_{10}$Kow value of 5.4 for dieldrin, determined using the "slow-stirring" method [46], was used as the input parameter for the BAF–QSAR model. Then, using Equation (2), TBC or TRV were converted to the WQC with the estimates of BAF, expressed as the concentration in the water.

$$\mathrm{WQC_{aqua}}(\mathrm{WQC_{wild}}) = \frac{\mathrm{TBC(TRV)}}{\mathrm{BAF}} \quad (2)$$

where $\mathrm{WQC_{aqua}}$ and $\mathrm{WQC_{wild}}$ represent the protection of aquatic life and aquatic-dependent wildlife, respectively. TBC is derived from tissue–based toxicity data using the SSD method (mg/kg), TRV is calculated using Equation (1), and BAF is the bioaccumulation factor (L/kg).

2.4. Ecological Risk Assessment

The available literature data on concentrations of dieldrin in China were used for environmental risk screening via the quotient method. Risk quotients (RQs) are the ratio of points estimates of exposure and toxicity. Exposure refers to actual monitored or model-estimated environment concentration. The toxicity refers to an effect level such as the predicted no-effect concentration (PNEC), which is the ecological risk threshold predicted to have no adverse effects on organisms. RQs were calculated as shown in Equation (3):

$$\mathrm{RQ} = \mathrm{C}/\mathrm{WQC} \quad (3)$$

where C is the measured environmental concentration collected from the literature; WQC for dieldrin is derived from Equation (2), which expresses the ecological risk threshold. Then, it is compared to the EPA's levels of concern (LOCs). According to the EPA's risk assumptions, the value of the LOC is 1 for chronic risk. A resulting RQ below the value of 1 indicates no chronic risk concern.

2.5. Data Analysis

The current study employs the R package "ssdtools" to generate the SSD [47] using R version 4.1.0 [48]. We utilized the "ssd_fit_dists" function using maximum likelihood to fit the distribution and the "ssd_gof" function to assess the goodness of fit. Using bootstrapping to obtain confidence intervals, the "ssd_hc" function was used to estimate the model-averaged 5% concentration; that is, the $\mathrm{HC_5}$ value. Together with the original data, the predictions were plotted using the "ssd_plot" function. The model BAF-QSAR v1.1 [45] coded in a Microsoft Excel workbook was used to obtain the BAF. The data screening and derivation of WQC were also carried out using Microsoft Excel 2019 for Windows.

3. Results and Discussion

3.1. WQC for the Protection of Aquatic Life

Table 1 presents toxicity data obtained from database for the dieldrin levels present in aquatic life. The selected species, including ten fishes, one crustacean and one mollusk, met the minimum data requirements. As shown in Figure 1, the SSD curve with the confidence interval was simulated from collated data. The TBC is 0.227 mg/kg (wet weight; WW), derived from estimated model-averaged predictions of 5% hazard concentrations with parametric bootstrapping.

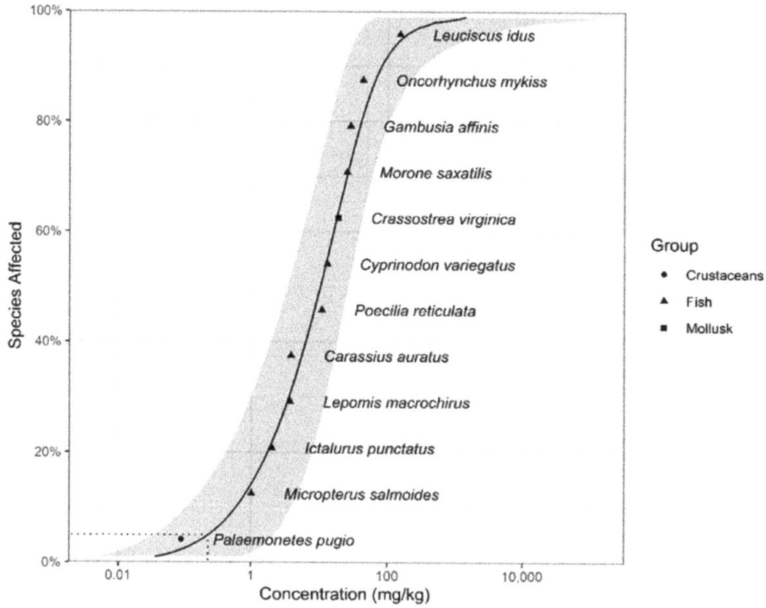

Figure 1. Species sensitivity distribution for dieldrin for aquatic organism tissue-based toxicity data. The model-averaged 95% confidence interval is indicated by the shaded band and the model-averaged 5% hazard concentration (HC5) by the dotted line.

Since data on dieldrin in the environment and biota are insufficient, the model approach was used to obtain the BAF. The estimation of the logarithm of BAFs to base 10 in the lower, middle and upper trophic levels of fish species are 4.77, 5.09 and, 5.66, respectively. The BAF predictions are considered generic, and they are not considered to be for a particular species of fish. Moreover, Arnot and Gobas [49] reviewed the status of bioaccumulation evaluations for organic chemicals in aquatic systems and summarized regression statistics for different organism classes before and after the confidence assessment on the reviewed data. In agreement with their empirical model, the $\log_{10}$BAF value of 4.77 was used to derive the WQC for dieldrin. The calculated BAF for fish is 58,884.37 L/kg. Using Equation (3), the calculated WQC for aquatic organisms is 3.86 ng/L.

Table 1. Tissue-based toxicity data of aquatic life for dieldrin.

Species	Common Name	NOEC (mg/kg, WW *)	Taxa	Reference
Palaemonetes pugio	Grass shrimp	0.09	Crustaceans	[50]
Micropterus salmoides	Largemouth bass	1.01	Fishes	[51]
Ictalurus punctatus	Channel catfish	2	Fishes	[52]
Lepomis macrochirus	Bluegill	3.7	Fishes	[53]
Carassius auratus	Goldfish	3.8	Fishes	[53]
Poecilia reticulata	Guppy	10.7	Fishes	[54]
Cyprinodon variegatus	Sheepshead minnow	12.8	Fishes	[50]
Crassostrea virginica	Eastern oyster	18.6	Mollusks	[55]
Morone saxatilis	Striped bass	25	Fishes	[56]
Gambusia affinis	Mosquito fish	28	Fishes	[57]
Oncorhynchus mykiss	Rainbow trout	43	Fishes	[58]
Leuciscus idus	Golden ide	151	Fishes	[59]

* WW = wet weight.

3.2. WQC for Protection of Aquatic-Dependent Wildlife

Table 2 presents the toxicity data for dieldrin of wildlife, including 11 avians and 10 mammalians which met the minimum data requirement. As shown in Figures 2 and 3, the SSD curves with confidence intervals were simulated for each taxa group from the collated data, and 5% hazard concentrations were calculated. The TDIs for avians and mammalians are 0.0363 and 0.0914 mg/kg per day, respectively. The highest FI:BW value among representative avian species in China is 0.43 [60], and the higher ratio of FI to BW among representative mammalian species in China is 0.5 [61–63]. The TRVs are 0.08442 and 0.1828 mg/kg WW, respectively. Using Equation (3) and the previous calculated BAF for fish, the calculated WQC are 1.4 and 3.10 ng/L, respectively. To better protect aquatic-dependent wildlife, the smallest WQC (1.40 ng/L) was selected as the dieldrin WQC for protecting aquatic-dependent wildlife in China.

Considering the bioaccumulation of dieldrin in piscivorous food webs, the threshold contaminant body burden in wildlife was calculated, which was then back-calculated to an equivalent concentration in fish or water using the combination of food and chemical assimilation efficiencies and bioconcentration/bioaccumulation factors. For wildlife criteria derived from dietary toxicity data, BAFs would be applied and appropriately weighted for each component of the aquatic diet of the representative wildlife species [42]. Using this method, the dieldrin wildlife value for waters within the Great Lakes Basin was determined to be 0.071 ng/L [23]. According to protocol used in the Canadian Tissue Residue Guidelines for the Protection of Wildlife that Consume Aquatic Biota [43], reference concentrations are calculated using information on BW and FI for wildlife species as well as the TDI derived from toxicity studies; thus, the result can be compared to the generic tissue residue guideline developed to protect all wildlife. For substances with a high potential for bio-magnification within food chains, it is important that the guideline used for lowest reference value is applied to the highest aquatic trophic level (e.g., level 4 fish) in order to protect predators feeding at that level.

Table 2. Tissue-based toxicity data of aquatic-dependent wildlife for dieldrin.

Species	Common Name	NOEC (mg/kg bw/day, WW) *	Taxa	Reference
Tyto alba	Barn owl	0.0445	Avian	[64]
Numida meleagris	Crowned guinea fowl	0.0537	Avian	[65]
Phasianus colchicus	Ring-necked pheasant	0.26	Avian	[66]
Colinus virginianus	Bobwhite quail	0.27	Avian	[67]
Streptopelia risoria	Ring dove	0.331	Avian	[68]
Coturnix japonica	Japanese quail	0.852	Avian	[69]
Phasianus colchicus	Pheasant	0.905	Avian	[70]
Coturnix coturnix	Quail	1.36	Avian	[71]
Columba livia	Homing pigeon	2	Avian	[72]
Anas platyrhynchos	Mallard	2.21	Avian	[73]
Gallus domesticus	Chicken	10	Avian	[74]
Canis familiaris	Dog	0.05	Mammalian	[75]
Mus musculus	Mouse	0.133	Mammalian	[76]
Damaliscus pygargus	Blesbuk	0.449	Mammalian	[77]
Odocoileus virginianus	White-tailed deer	0.72	Mammalian	[78]
Rattus novegicus	Rat	0.81	Mammalian	[75]
Ovis aries	Sheep	0.87	Mammalian	[79]
Peromyscus leucopus	White-footed mouse	1.14	Mammalian	[80]
Sus scrofa	Pig	1.20	Mammalian	[81]
Oryctolagus cuniculus	Rabbit	1.25	Mammalian	[82]
Cavia porcellus	Guinea pig	1.76	Mammalian	[83]

* bw = body weight; WW = wet weight.

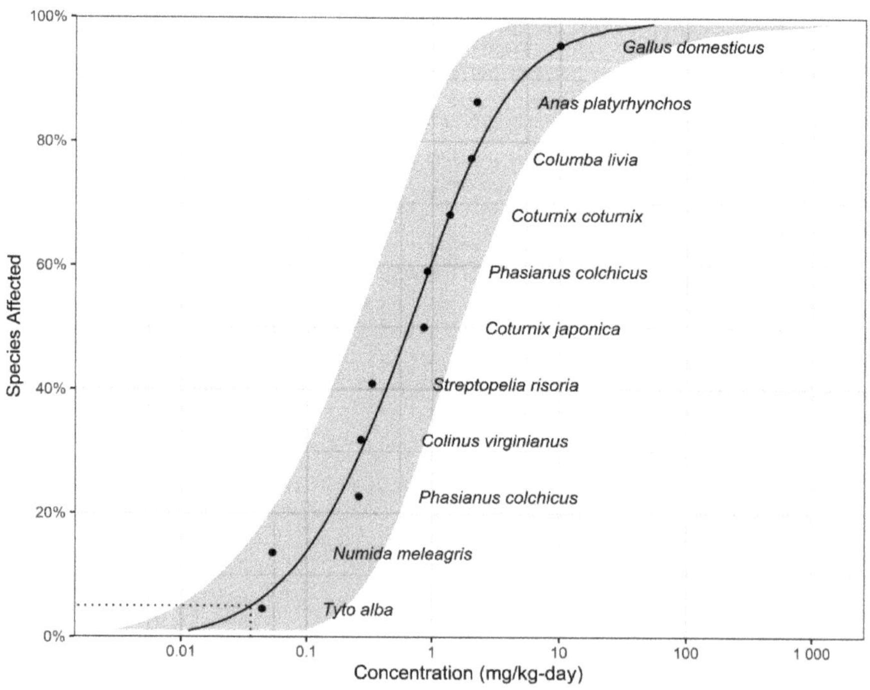

Figure 2. Species sensitivity distribution for dieldrin for avian tissue-based toxicity data. The model-averaged 95% confidence interval is indicated by the shaded band and the model-averaged 5% hazard concentration (HC5) by the dotted line.

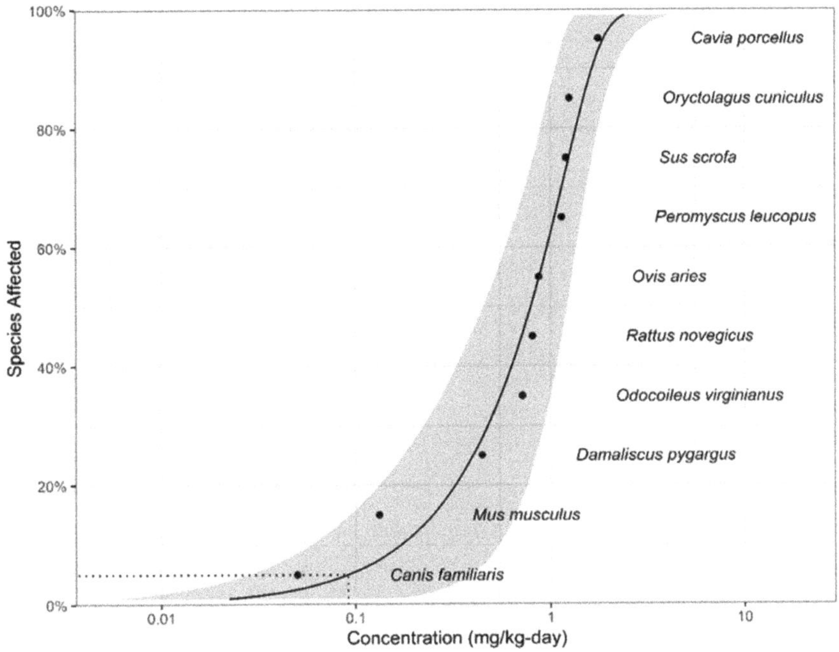

Figure 3. Species sensitivity distribution for dieldrin for mammalian tissue-based toxicity data. The model-averaged 95% confidence interval is indicated by the shaded band and the model-averaged 5% hazard concentration (HC5) by the dotted line.

Since species-specific and site-specific data for dieldrin were not available in China, we selected the highest ratio of FI to BW of representative wildlife species to calculate the TRV from TDI and used BAF for fish to convert the TRV to water concentration. Dieldrin was found in a fish liver sample (0.07 μg/g WW) collected from aquaculture cages in coastal waters of Xiamen [84] and in various organs and tissues of Indo-Pacific humpback dolphins from the Pearl River Estuary, with values ranging from 0.74 to 6.8 ng/g WW [28]. Our WQC is much higher than that of the Great Lakes Basin, but compared to residue records in biota, it should be effective for protecting wildlife.

3.3. Risk Assessment

Studies have shown that chronic exposure to dieldrin can cause harm to aquatic organisms and wildlife [10,85]. Therefore, it is important to assess the ecological risk of dieldrin in aquatic environments. The data on the presence of dieldrin in surface water from the last dozen years, identified by the literature review and the ecological risk assessment, are shown in Table 3. Samples were collected from the Yangtze River, the Qinhuai River and the Xuanwu Lake in Nanjing [27] and the Shaying River Basin [86], and the mean concentrations of dieldrin were 1.31, 2.32, 4.38 and 4.6 ng/L, respectively. RQ values were calculated separately using WQC for the protection of aquatic life and wildlife. Results ranging between 0.34 and 3.29 were observed, suggesting that dieldrin poses an ecological risk to both aquatic organism and wildlife in the Xuanwu Lake and the Shaying River Basin, poses a potential ecological risk to avians in the Qinhuai River and poses no chronic risks in the Yangtze River.

Table 3. RQ estimations.

Location	Conc/(ng/L)	RQ_Aqua	RQ_Wild	Sampling Time	Reference
Yangtze River, Nanjing	1.31	0.34	0.94	May to July 2016	[27]
Qinhuai River, Nanjing	2.32	0.60	1.66		
Xuanwu Lake, Nanjing	4.38	1.13	3.13		
Shaying River Basin	4.60	1.19	3.29	November 2013	[86]

Regarding the sampling approach, as the timing and site selection were unable to adequately capture the periodic occurrence of pesticides or investigate surface waters particularly susceptible to pesticide risks, this may contribute to an inappropriate estimation of risk [87]. Since dieldrin has been banned and is not used in China, its main source is environment migration. With hydrophobicity and low water solubility, dieldrin is prone to binding to organic materials [1]. Research on the distribution of organochlorine pesticide pollution in Indonesia revealed that sediments showed higher organochlorine concentrations than water, mollusks or fish [15]. Sediment is still a potential source and is rarely detected. It has been demonstrated that the concentration in sediments can be accurately predicted by multiplying the concentration in water by the chemical's organic carbon partition coefficient [22]. Thus, detecting dieldrin concentrations in water is important for better understanding its ecological risk to integral aquatic ecosystems.

4. Conclusions

This study set out to determine the WQC for dieldrin using tissue-based toxicity data. In this study, the estimated BAF of 58,884.37 L/kg was used to obtain the WQC. The tissue-based criteria obtained using screened dieldrin toxicity data and the SSD method are 0.227 mg/kg (WW) for aquatic life and 0.08442 and 0.1828 mg/kg of food per day (WW) for avian and mammalian species, respectively. Using the statistics presented above, the dieldrin WQC needed for the protection of aquatic organisms and wildlife in China are 3.86 and 1.4 ng/L, respectively. This study is limited by the lack of information on toxicity data and BAF measurements of dieldrin. Notwithstanding these limitations, this work contributes to furthering the protection of aquatic organisms and wildlife from the bioaccumulation of dieldrin, supporting environmental management and risk assessment in China. The results also enhance China's water environment criterion system, enabling the development of an "ecological water civilization". This research has also shown that dieldrin levels in surface water in parts of China may pose potential ecological risks to aquatic organisms and wildlife, especially avian species. These findings are cause for concern of the presence of dieldrin residue in Chinese aqueous environments. Meanwhile, considerably more work needs to be carried out to monitor dieldrin levels in surface water environments in China.

Author Contributions: Conceptualization, L.X. and H.S.; Data curation, L.X. and L.B.; Formal analysis, L.B. and Y.Z.; Funding acquisition, H.S.; Investigation, L.X.; Methodology, X.L. (Xuemei Li) and H.S.; Supervision, H.S.; Validation, X.L. (Xuesong Liu), F.W., Y.W., N.J. and M.Z.; Visualization, L.X. and L.B.; Writing—original draft, L.X.; Writing—review and editing, L.X., X.L. (Xuemei Li) and H.S. All authors have read and agreed to the published version of the manuscript.

Funding: This research was funded by the National Key Research and Development Program of China, grant number 2022YFC3702403, and National Natural Science Foundation of China, grant number 42394154.

Institutional Review Board Statement: Not applicable.

Informed Consent Statement: Not applicable.

Data Availability Statement: The original contributions presented in this study are included in the article. Further inquiries can be directed to the corresponding author.

Conflicts of Interest: The authors declare no conflicts of interest.

References

1. Boland, M.P. Dieldrin. In *Encyclopedia of Toxicology*, 4th ed.; Wexler, P., Ed.; Academic Press: Oxford, UK, 2024; pp. 685–689.
2. Bus, J.S.; Leber, A.P. Miscellaneous Chlorinated Hydrocarbon Pesticides. In *Patty's Toxicology*; Wiley: Hoboken, NJ, USA, 2001.
3. ASTDR. *Toxicological Profile for Aldrin and Dieldrin*; Agency for Toxic Substances and Disease Registry: Atlanta, GA, USA, 2022.
4. Petrocelli, S.R.; Anderson, J.W.; Hanks, A.R. Controlled food-chain transfer of dieldrin residues from phytoplankters to clams. *Mar. Biol.* **1975**, *31*, 215–218. [CrossRef]
5. Petrocelli, S.R.; Anderson, J.W.; Hanks, A.R. Biomagnification of dieldrin residues by food-chain transfer from clams to blue crabs under controlled conditions. *Bull. Environ. Contam. Toxicol.* **1975**, *13*, 108–116. [CrossRef]
6. Yamada, H.; Jandric, Z.; Chhem-Kieth, S.; Vreysen, M.J.; Rathor, M.N.; Gilles, J.R.; Cannavan, A. Anopheles arabiensis egg treatment with dieldrin for sex separation leaves residues in male adult mosquitoes that can bioaccumulate in goldfish (*Carassius auratus auratus*). *Environ. Toxicol. Chem.* **2013**, *32*, 2786–2791. [CrossRef] [PubMed]
7. Stern, A.H. Hazard identification of the potential for dieldrin carcinogenicity to humans. *Environ. Res.* **2014**, *131*, 188–214. [CrossRef] [PubMed]
8. Martyniuk, C.J.; Mehinto, A.C.; Denslow, N.D. Organochlorine pesticides: Agrochemicals with potent endocrine-disrupting properties in fish. *Mol. Cell. Endocrinol.* **2020**, *507*. [CrossRef] [PubMed]
9. Evangelista de Duffard, A.M.; Duffard, R. Behavioral toxicology, risk assessment, and chlorinated hydrocarbons. *Env. Health Perspect* **1996**, *104* (Suppl. 2), 353–360. [CrossRef] [PubMed]
10. Deck, A.T.; Reinke, E.N.; McCain, W.C. Wildlife Toxicity Assessment for Aldrin and Dieldrin. In *Wildlife Toxicity Assessments for Chemicals of Military Concern*; Elsevier: Amsterdam, The Netherlands, 2015; pp. 367–384.
11. Ulugov, U.A.; Bobritskaya, L.S.; Sinitsky, J. Inventory of Obsolete Pesticide Warehouses in Tajikistan and Implications for Removal of Contaminated Soil. *J. Health Pollut.* **2018**, *8*, 1–5. [CrossRef] [PubMed]
12. Navarrete, I.A.; Tee, K.A.M.; Unson, J.R.S.; Hallare, A.V. Organochlorine pesticide residues in surface water and groundwater along Pampanga River, Philippines. *Environ. Monit. Assess.* **2018**, *190*, 289. [CrossRef] [PubMed]
13. Ogbeide, O.; Uhunamure, G.; Okundaye, F.; Ejeomo, C. First report on probabilistic risk assessment of pesticide residues in a riverine ecosystem in South-South Nigeria. *Chemosphere* **2019**, *231*, 546–561. [CrossRef] [PubMed]
14. Snow, D.D.; Chakraborty, P.; Uralbekov, B.; Satybaldiev, B.; Sallach, J.B.; Thornton Hampton, L.M.; Jeffries, M.; Kolok, A.S.; Bartelt-Hunt, S.B. Legacy and current pesticide residues in Syr Darya, Kazakhstan: Contamination status, seasonal variation and preliminary ecological risk assessment. *Water Res.* **2020**, *184*, 116141. [CrossRef]
15. Oginawati, K.; Susetyo, S.H.; Rahmawati, S.I.; Kurniawan, S.B.; Abdullah, S.R.S. Distribution of organochlorine pesticide pollution in water, sediment, mollusk, and fish at Saguling Dam, West Java, Indonesia. *Toxicol. Res.* **2022**, *38*, 149–157. [CrossRef] [PubMed]
16. Souza, R.C.; Portella, R.B.; Almeida, P.; Pinto, C.O.; Gubert, P.; Santos da Silva, J.D.; Nakamura, T.C.; do Rego, E.L. Human milk contamination by nine organochlorine pesticide residues (OCPs). *J. Environ. Sci. Health Part B* **2020**, *55*, 530–538. [CrossRef] [PubMed]
17. El-Saeid, M.H.; Hassanin, A.S.; Bazeyad, A.Y. Levels of pesticide residues in breast milk and the associated risk assessment. *Saudi J. Biol. Sci.* **2021**, *28*, 3741–3744. [CrossRef] [PubMed]
18. Al Antary, T.M.; Alawi, M.A.; Kiwan, R.; Haddad, N.A. Monitoring of Organochlorine Pesticide Residues in Human Breast Milk in the Northern Governorates of Jordan in 2019/2020 Compared with the Results of 2015 Study. *Bull. Environ. Contam. Toxicol.* **2021**, *106*, 1071–1076. [CrossRef] [PubMed]
19. Padayachee, K.; Reynolds, C.; Mateo, R.; Amar, A. A global review of the temporal and spatial patterns of DDT and dieldrin monitoring in raptors. *Sci. Total Environ.* **2023**, *858*, 159734. [CrossRef] [PubMed]
20. Pedersen, A.F.; Dietz, R.; Sonne, C.; Letcher, R.J.; Roos, A.M.; Simon, M.; Rosing-Asvid, A.; Ferguson, S.H.; McKinney, M.A. Feeding and biological differences induce wide variation in legacy persistent organic pollutant concentrations among toothed whales and polar bear in the Arctic. *Sci. Total Environ.* **2024**, *908*. [CrossRef] [PubMed]

21. CCME. *Canadian Sediment Quality Guidelines for the Protection of Aquatic Life-dieldrin*; Canadian Council of Ministers of the Environment: Ottawa, ON, Canada, 1999.
22. USEPA. *Procedures for the Derivation of Equilibrium Partitioning Sediment Benchmarks (ESBs) for the Protection of Benthic Organisms: Dieldrin*; Office of Research and Development: Washington, DC, USA, 2003.
23. USEPA. *Wildlife Fact Sheet for Dieldrin: (Wildlife)*; Indiana Department of Environmental Management: Indianapolis, IN, USA, 1998.
24. Meharg, A.; Wright, J.; Osborn, D. The frequency of Environmental Quality Standard (EQS) exceedance for chlorinated organic pollutants in rivers of the Humber catchments. *Sci. Total Environ.* **1998**, *210–211*, 219–228. [CrossRef]
25. Niu, L.; Xu, C.; Zhu, S.; Liu, W. Residue patterns of currently, historically and never-used organochlorine pesticides in agricultural soils across China and associated health risks. *Environ. Pollut.* **2016**, *219*, 315–322. [CrossRef] [PubMed]
26. Wang, L.; Cao, G.; Liu, L.-Y.; Zhang, Z.-F.; Jia, S.-M.; Fu, M.-Q.; Ma, W.-L. Cross-regional scale studies of organochlorine pesticides in air in China: Pollution characteristic, seasonal variation, and gas/particle partitioning. *Sci. Total Environ.* **2023**, *904*, 166709. [CrossRef] [PubMed]
27. Chen, Y.; Huang, R.; Guan, Y.; Zhuang, T.; Wang, Y.; Tan, R.; Wang, J.; Zhou, R.; Wang, B.; Xu, J.; et al. The profiling of elements and pesticides in surface water in Nanjing, China with global comparisons. *Sci. Total Environ.* **2021**, *774*, 145749. [CrossRef]
28. Zhang, X.; Zhan, F.; Yu, R.-Q.; Sun, X.; Wu, Y. Bioaccumulation of legacy organic contaminants in pregnant Indo-Pacific humpback dolphins (Sousa chinensis): Unique features on the transplacental transfer. *Sci. Total Environ.* **2021**, *785*, 147287. [CrossRef] [PubMed]
29. Zhao, Z.; Zhang, L.; Wu, J.; Fan, C. Residual levels, tissue distribution and risk assessment of organochlorine pesticides (OCPs) in edible fishes from Taihu Lake, China. *Environ. Monit. Assess.* **2013**, *185*, 9265–9277. [CrossRef]
30. Zhang, A.; Fang, L.; Wang, J.; Liu, W.; Yuan, H.; Jantunen, L.; Li, Y.-F. Residues of Currently and Never Used Organochlorine Pesticides in Agricultural Soils from Zhejiang Province, China. *J. Agric. Food Chem.* **2012**, *60*, 2982–2988. [CrossRef]
31. Han, X.; Xu, L.; Deng, A.; Xing, P.; Xu, Y. Centurial deposition records of polychlorinated biphenyls and organochlorine pesticides in sediment cores from a plateau deep-water lake of China: Significance of anthropogenic impacts, transformation signals and ecological risks revealed by full congener analysis. *Sci. Total Environ.* **2024**, *926*. [CrossRef]
32. *GB3838-2002*; Environmental Quality Standard for Surface Water. State Environmental Protection Administration: Beijing, China, 2002; pp. 1–8.
33. Yang, M.; Mu, Y.; Wu, F. Derivation on Criteria Maximum Concentration of Four Typical Organochlorine Pesticides for Protecting Aquatic Organisms. *J. Kunming Univ. Sci. Technol. (Nat. Sci.)* **2019**, *44*, 111–118. [CrossRef]
34. Barnhart, B.; Flinders, C.; Ragsdale, R.; Johnson, G.; Wiegand, P. Deriving Human Health and Aquatic Life Water Quality Criteria in the United States for Bioaccumulative Substances: A Historical Review and Future Perspective. *Environ. Toxicol. Chem.* **2021**, *40*, 2394–2405. [CrossRef]
35. Meador, J.P.; Warne, M.S.J.; Chapman, P.M.; Chan, K.M.; Yu, S.; Leung, K.M.Y. Tissue-based environmental quality benchmarks and standards. *Environ. Sci. Pollut. Res.* **2013**, *21*, 28–32. [CrossRef]
36. Hamilton, S.J. Rationale for a tissue-based selenium criterion for aquatic life. *Aquat. Toxicol.* **2002**, *57*, 85–100. [CrossRef] [PubMed]
37. Beckvar, N.; Dillon, T.M.; Read, L.B. Approaches for linking whole-body fish tissue residues of mercury or DDT to biological effects thresholds. *Environ. Toxicol. Chem.* **2005**, *24*, 2094–2105. [CrossRef] [PubMed]
38. Steevens, J.A.; Reiss, M.R.; Pawlisz, A.V. A methodology for deriving tissue residue benchmarks for aquatic biota: A case study for fish exposed to 2,3,7,8-tetrachlorodibenzo-p-dioxin and equivalents. *Integr. Environ. Assess. Manag.* **2005**, *1*, 142–151. [CrossRef] [PubMed]
39. Meador, J. Rationale and Procedures for Using the Tissue-Residue Approach for Toxicity Assessment and Determination of Tissue, Water, and Sediment Quality Guidelines for Aquatic Organisms. *Hum. Ecol. Risk Assess. Int. J.* **2006**, *12*, 1018–1073. [CrossRef]
40. Fox, D.R.; van Dam, R.A.; Fisher, R.; Batley, G.E.; Tillmanns, A.R.; Thorley, J.; Schwarz, C.J.; Spry, D.J.; McTavish, K. Recent Developments in Species Sensitivity Distribution Modeling. *Environ. Toxicol. Chem.* **2021**, *40*, 293–308. [CrossRef] [PubMed]
41. Wheeler, J.R.; Grist, E.P.M.; Leung, K.M.Y.; Morritt, D.; Crane, M. Species sensitivity distributions: Data and model choice. *Mar. Pollut. Bull.* **2002**, *45*, 192–202. [CrossRef]
42. USEPA. *Tissue-Based Criteria for "Bioaccumulative" Chemicals*; USEPA: Washington, DC, USA, 2005.
43. CCME. *Protocol for the Derivation of Canadian Tissue Residue Guidelines for the Protection of Wildlife that Consume Aquatic Biota*; Canadian Council of Ministers of the Environment: Winnipeg, MB, USA, 1998.
44. Mackay, D.; Fraser, A. Kenneth Mellanby Review Award. Bioaccumulation of persistent organic chemicals: Mechanisms and models. *Environ. Pollut.* **2000**, *110*, 375–391. [CrossRef]
45. Arnot, J.A.; Gobas, F.A.P.C. A Generic QSAR for Assessing the Bioaccumulation Potential of Organic Chemicals in Aquatic Food Webs. *QSAR Comb. Sci.* **2003**, *22*, 337–345. [CrossRef]
46. De Bruijn, J.; Busser, F.; Seinen, W.; Hermens, J. Determination of octanol/water partition coefficients for hydrophobic organic chemicals with the "slow-stirring" method. *Environ. Toxicol. Chem.* **1989**, *8*, 499–512. [CrossRef]

47. Thorley, J.; Schwarz, C. ssdtools: An R package to fit Species Sensitivity Distributions. *J. Open Source Softw.* **2018**, *3*. [CrossRef]
48. R Core Team. *R: A Language and Environment for Statistical Computing, 4.1.0*; R Foundation for Statistical Computing: Vienna, Austria, 2021.
49. Arnot, J.A.; Gobas, F.A.P.C. A review of bioconcentration factor (BCF) and bioaccumulation factor (BAF) assessments for organic chemicals in aquatic organisms. *Environ. Rev.* **2006**, *14*, 257–297. [CrossRef]
50. Parrish, P.R.; Couch, J.A.; Forester, J.; Patrick, J.M., Jr.; Cook, G.H. Dieldrin: Effects on Several Estuarine Organisms. *J. Southeast. Assoc. Fish Wildl. Agencies* **1977**, 427–434.
51. Johnson, K.G.; Muller, J.K.; Price, B.; Ware, A.; Sepúlveda, M.S.; Borgert, C.J.; Gross, T.S. Influence of seasonality and exposure on the accumulation and reproductive effects of p,p′-dichlorodiphenyldichloroethane and dieldrin in largemouth bass. *Environ. Toxicol. Chem.* **2007**, *26*, 927–934. [CrossRef]
52. Shannon, L.R. Accumulation and elimination of dieldrin in muscle tissue of channel catfish. *Bull. Environ. Contam. Toxicol.* **1977**, *17*, 637–644. [CrossRef]
53. Gakstatter, J.H.; Weiss, C.M. The Elimination of DDT-C14, Dieldrin-C14, and Lindane-C14 from Fish following a Single Sublethal Exposure in Aquaria. *Trans. Am. Fish. Soc.* **1967**, *96*, 301–307. [CrossRef]
54. Burnett, K.M.; Liss, W.J. Multi-steady-state toxicant fate and effect in laboratory aquatic ecosystems. *Environ. Toxicol. Chem.* **1990**, *9*, 637–647. [CrossRef]
55. Mason, J.W.; Rowe, D.R. The accumulation and loss of dieldrin and endrin in the Eastern oyster. *Arch. Environ. Contam. Toxicol.* **1976**, *4*, 349–360. [CrossRef]
56. Santerre, C.R.; Blazer, V.S.; Khanna, N.; Reinert, R.E.; Barrows, F.T. Absorption of Dietary Dieldrin by Striped Bass. *Bull. Environ. Contam. Toxicol.* **1997**, *58*, 334–340. [CrossRef]
57. Metcalf, R.L. Chapter 2—A Laboratory Model Ecosystem to Evaluate Compounds Producing Biological Magnification. In *Essays in Toxicology*; Hayes, W.J., Ed.; Elsevier: Amsterdam, The Netherlands, 1974; Volume 5, pp. 17–38.
58. Mehrle, P.; Mayer, F.; Johnson, W.; Mayer, F.; Hamelink, J. Diet Quality in Fish Toxicology: Effects on Acute and Chronic Toxicity. In *Aquatic Toxicology and Hazard Evaluation*; ASTM International: West Conshohocken, PA, USA, 1977.
59. Freitag, D.; Ballhorn, L.; Geyer, H.; Korte, F. Environmental hazard profile of organic chemicals: An experimental method for the assessment of the behaviour of organic chemicals in the ecosphere by means of simple laboratory tests with 14C labelled chemicals. *Chemosphere* **1985**, *14*, 1589–1616. [CrossRef]
60. Su, H.; Mu, Y.; Feng, C.; Zhu, Y.; Wang, H.; Wu, F.; Giesy, J.P. Tissue Residue Guideline for ∑DDT for Protection of Aquatic Birds in China. *Hum. Ecol. Risk Assess. Int. J.* **2014**, *20*, 1629–1642. [CrossRef]
61. Su, H.; Feng, C.; Chang, H.; Mu, Y.; Wu, F. Study on Tissue Residue Guidelines of DDTs for Protection of Aquatic Mammalian Species in China. *Asian J. Ecotoxicol.* **2015**, *10*, 110–118.
62. Hung, C.L.H.; Lau, R.K.F.; Lam, J.C.W.; Jefferson, T.A.; Hung, S.K.; Lam, M.H.W.; Lam, P.K.S. Risk assessment of trace elements in the stomach contents of Indo-Pacific Humpback Dolphins and Finless Porpoises in Hong Kong waters. *Chemosphere* **2007**, *66*, 1175–1182. [CrossRef] [PubMed]
63. Jefferson, T.A. Population biology of the Indo-Pacific Humpacked dolphin in Hong Kong waters. *J. Wildl. Manag.* **2000**, *64*, 1–65.
64. Mendenhall, V.M.; Klaas, E.E.; McLane, M.A.R. Breeding success of barn owls (*Tyto alba*) fed low levels of DDE and dieldrin. *Arch. Environ. Contam. Toxicol.* **1983**, *12*, 235–240. [CrossRef]
65. Wiese, I.H.; Basson, N.C.J.; Van Der Vyver, J.H.; Van Der Merwe, J.H. Toxicology and Dynamics of Dieldrin in the Crowned Guinea-fowl, Numida Meleagris (L). *Phytophylactica* **1969**, *1*, 161–175.
66. Atkins, T.D. Effects of Dieldrin on Reproduction of Penned Hen Pheasants. *J. Wildl. Manag.* **1967**, *31*, 746. [CrossRef]
67. Fergin, T.J.; Schafer, E.C. Toxicity of dieldrin to bobwhite quail in relation to sex and reproductive status. *Arch. Environ. Contam. Toxicol.* **1977**, *6*, 213–219. [CrossRef] [PubMed]
68. Heinz, G.H.; Hill, E.F.; Contrera, J.F. Dopamine and norepinephrine depletion in ring doves fed DDE, dieldrin, and Aroclor 1254. *Toxicol. Appl. Pharmacol.* **1980**, *53*, 75–82. [CrossRef] [PubMed]
69. Reading, C.M.; Arscott, G.H.; Tinsley, I.J. Effect of Dieldrin and Calcium on the Performance of Adult Japanese Quail (*Coturnix coturnix japonica*). *Poult. Sci.* **1976**, *55*, 212–219. [CrossRef]
70. Dahlgren, R.B.; Linder, R.L. Effects of Dieldrin in Penned Pheasants through the Third Generation. *J. Wildl. Manag.* **1974**, *38*, 320. [CrossRef]
71. Kreitzer, J.F.; Heinz, G.H. The effect of sublethal dosages of five pesticides and a polychlorinated biphenyl on the avoidance response of coturnix quail chicks. *Environ. Pollut. (1970)* **1974**, *6*, 21–29. [CrossRef]
72. Jefferies, D.J.; French, M.C. Changes Induced in the Pigeon Thyroid by p,p prime -DDE and Dieldrin. *J. Wildl. Manag.* **1972**, *36*, 24. [CrossRef]
73. Davison, K.L.; Sell, J.L. DDT thins shells of eggs from mallard ducks maintained onad libitum or controlled-feeding regimens. *Arch. Environ. Contam. Toxicol.* **1974**, *2*, 222–232. [CrossRef] [PubMed]
74. Eden, W.G. Toxicity of Dieldrin to Chickens. *J. Econ. Entomol.* **1951**, *44*, 1013. [CrossRef]

75. Walker, A.I.T.; Stevenson, D.E.; Robinson, J.; Thorpe, E.; Roberts, M. The toxicology and pharmacodynamics of dieldrin (HEOD): Two-year oral exposures of rats and dogs. *Toxicol. Appl. Pharmacol.* **1969**, *15*, 345–373. [CrossRef] [PubMed]
76. Walker, A.I.T.; Thorpe, E.; Stevenson, D.E. The toxicology of dieldrin (HEOD). I. Long-term oral toxicity studies in mice. *Food Cosmet. Toxicol.* **1973**, *11*, 415–432. [CrossRef] [PubMed]
77. Wiese, I.; Basson, N.C.J.; Basson, P.; Naudé, T.W.; Maartens, B.P. The toxicology and pathology of dieldrin and photodieldrin poisoning in two antelope species. *Onderstepoort J. Vet. Res.* **1973**, *40 1*, 31–39.
78. Murphy, D.A.; Korschgen, L.J. Reproduction, growth, and tissue residues of deer fed dieldrin. *J. Wildl. Manag.* **1970**, *34*, 887–903. [CrossRef]
79. Davison, K.L. Growth, Hemoglobin, Body Composition and Vitamin A of Sheep Fed Dieldrin. *J. Anim. Sci.* **1970**, *31*, 567–575. [CrossRef]
80. Bildstein, K.L.; Forsyth, D.J. Effects of dietary dieldrin on behavior of white-footed mice (*Peromyscus leucopus*) towards an avian predator. *Bull. Environ. Contam. Toxicol.* **1979**, *21*, 93–97. [CrossRef]
81. Kotonya, R.; Jensen, N.-E. No effect of dieldrin on progesterone production in gilts. *Toxicology* **1993**, *81*, 165–171. [CrossRef] [PubMed]
82. Hurkat, P.C.; Joshi, G.P. Some Physiological and Behavior Studies on Rabbit (*Oryctolagus cuniculus*) after Dieldrin Administration. *Indian Vet. J.* **1977**, *54*, 709–714.
83. Uzoukwu, M.; Sd, S. Dieldrin toxicosis: Fetotoxicosis, tissue concentrations, and microscopic and ultrastructural changes in guinea pigs. *Am. J. Vet. Res.* **1972**, *33 3*, 579–583.
84. Klump, D.W.; Huasheng, H.; Humphrey, C.; Xinhong, W.; Codi, S. Toxic contaminants and their biological effects in coastal waters of Xiamen, China. I. Organic pollutants in mussel and fish tissues. *Mar. Pollut. Bull.* **2002**, *44*, 752–760. [CrossRef] [PubMed]
85. Lamai, S.L.; Warner, G.F.S.; Walker, C.H. Effects of dieldrin on life stages of the African catfish, *Clarias gariepinus* (Burchell). *Ecotoxicol. Environ. Saf.* **1999**, *42 1*, 22–29. [CrossRef]
86. Bai, Y.; Ruan, X.; van der Hoek, J.P. Residues of organochlorine pesticides (OCPs) in aquatic environment and risk assessment along Shaying River, China. *Environ. Geochem. Health* **2018**, *40*, 2525–2538. [CrossRef]
87. Weisner, O.; Arle, J.; Liebmann, L.; Link, M.; Schafer, R.B.; Schneeweiss, A.; Schreiner, V.C.; Vormeier, P.; Liess, M. Three reasons why the Water Framework Directive (WFD) fails to identify pesticide risks. *Water Res.* **2022**, *208*, 117848. [CrossRef] [PubMed]

toxics

MDPI

Article

Acute Toxicity, Neurotoxic, Immunotoxic, and Behavioral Effects of Deltamethrin and Sulfamethoxazole in Adult Zebrafish: Insights into Chemical Interactions and Environmental Implications

Yueyue Liu [1,2], Fengyu Liu [3] and Chen Wang [2,*]

[1] College of Water Science, Beijing Normal University, Beijing 100875, China; liuyueyue0103@mail.bnu.edu.cn
[2] State Key Laboratory of Environmental Criteria and Risk Assessment, Chinese Research Academy of Environmental Sciences, Beijing 100012, China
[3] State Environment Protection Key Laboratory of Environmental Monitoring Quality Control, China National Environmental Monitoring Center, Beijing 100012, China; liufengyu@cnemc.cn
* Correspondence: wangchen@craes.org.cn

Abstract: The increasing presence of antimicrobial agents and pesticides in aquatic environments raises concerns about their potential impacts on non-target organisms. Among these chemicals, deltamethrin (DM), a widely used pesticide, and sulfamethoxazole (SMX), an antimicrobial commonly detected in water bodies, pose significant ecological risks. This study investigates the acute toxicity, neurotoxic effects, oxidative stress responses, immune-related gene expression, and feeding behavior of adult zebrafish exposed to DM and SMX. The 96 h LC_{50} for DM was 4.84 μg/L, indicating significant acute toxicity, while the LC_{50} for the DM + SMX mixture was 11.32 μg/L, suggesting that SMX may mitigate the toxicity of DM. Neurotransmitter alterations, including reduced levels of γ-aminobutyric acid (γ-GABA), serotonin (5-HT), and acetylcholinesterase (AChE), were observed, with the combination of DM and SMX showing partial restoration of AChE activity. Oxidative stress markers revealed significant changes in antioxidant enzyme activities, with DM exposure increasing superoxide dismutase (SOD) and glutathione-S-transferase (GST) activities, while decreasing catalase (CAT) and glutathione peroxidase (GPX) activities. Immune-related gene expression demonstrated suppressed IgM, IgD, and IgZ levels, along with altered inflammatory responses, with both DM and DM + SMX exposure inducing pro-inflammatory cytokines. Finally, feeding behavior was significantly impaired in the DM group at the 3 min mark, while the DM + SMX group showed partial mitigation of this effect. These findings highlight the neurotoxic, immunotoxic, and behavioral effects of DM and SMX, and underscore the potential for chemical interactions to modulate toxicity in aquatic organisms.

Keywords: pyrethroids; antibiotic; immunotoxicity; neurotoxicity; zebrafish

1. Introduction

Deltamethrin (DM), a widely used pyrethroid insecticide, has raised significant concerns due to its neurotoxic effects on aquatic organisms, including fish [1,2]. Pyrethroids like DM exert their toxicity primarily by disrupting sodium ion channels in the nervous system, which leads to excessive neural stimulation and, ultimately, mortality in aquatic species [3]. As a commonly applied pesticide, DM is frequently detected in aquatic ecosystems, such as rivers, lakes, and streams, particularly those affected by runoff from agricultural fields, where it becomes a major environmental contaminant [4,5]. The concentrations of DM are reported to range from ng/L to μg/L [6–9], which are high enough to pose a risk to aquatic organisms, especially in the case of chronic exposure or during periods of heavy runoff. Sulfamethoxazole (SMX) is a widely used synthetic antimicrobial agent belonging to the sulfonamide class of antibiotics [10]. It is commonly widely used in both human medicine and aquaculture to treat a variety of bacterial infections [11,12]. SMX is also frequently detected in aquatic environments due to its widespread use and limited degradation in natural water bodies [11,13]. As another pervasive contaminant in water systems, SMX poses a potential threat to aquatic organisms, as it can accumulate in water and affect the health of aquatic species, including fish [12,14].

The environmental persistence of SMX, combined with its frequent co-occurrence with other pollutants, such as pesticides, raises concerns about the potential synergistic or antagonistic effects of chemical mixtures on aquatic organisms. Recent studies indicated that mixtures of environmental contaminants can alter the toxicity of individual chemicals through various mechanisms, including metabolic interactions and changes in bioavailability [15–17]. The toxicological effects of individual chemicals have been extensively studied. However, there is limited research on the combined toxicity of pesticides and antibiotics, particularly concerning neurotoxicity, oxidative stress, and immune dysfunction in aquatic species. Given the frequent detection of these chemicals together in aquatic environments, it is essential to explore how their combination impacts the health of aquatic organisms.

Zebrafish (Danio rerio) have emerged as a prominent model organism for toxicological studies due to their well-documented sensitivity to environmental contaminants and their physiological similarity to other vertebrates, including humans [18,19]. This study investigates the acute toxicity of DM, both individually and in combination with SMX, in adult male zebrafish to explore the potential neurotoxic, immunotoxic, and behavioral effects of these chemicals. Altered feeding behavior often signals disruptions in neurological function, energy metabolism, or general health [20]. Given that pyrethroids like DM are known to affect neural pathways, we hypothesized that exposure to DM may impair zebrafish feeding behavior. Furthermore, we aimed to evaluate whether the presence of SMX modulates these effects. In addition to behavioral changes, we also investigated the effects of these chemicals on neurotransmitter levels, immune-related gene expression, and oxidative stress markers in the brain. These biomarkers are crucial for understanding the molecular mechanisms underlying toxicity, as they provide insights into the disruption of neurochemical signaling, immune function, and cellular stress responses, which may have long-term implications for fish populations and aquatic ecosystems. The neurotransmitters γ-aminobutyric acid (GABA), serotonin (5-HT), and acetylcholinesterase (AChE) are key indicators of neurotoxicity in zebrafish [21], with alterations in their levels reflecting disturbances in neural signaling [22,23]. Moreover, oxidative stress and immune dysfunction, which are often induced by chemical exposure, can contribute to long-term neurotoxic effects [24–26].

Overall, this study seeks to address this gap by investigating the acute toxicity, neurotoxic effects, and immune-related impacts of DM, both individually and in combination with SMX, in adult zebrafish. Specifically, we aim to examine alterations in neurotransmitter alterations, oxidative stress markers, and immune function, all of which are key indicators of chemical-induced neurotoxicity and systemic disruption. By investigating these endpoints, we aim to better understand the potential risks posed by the co-occurrence of pesticides and antibiotics in aquatic environments and the implications for the health of aquatic organisms and inform future risk assessment efforts for aquatic ecosystems.

2. Materials and Methods

2.1. Chemicals and Reagents

Deltamethrin (≥98% purity), sulfamethoxazole (99% purity), and dimethyl sulfoxide (DMSO, chromatographic grade) were purchased from Sigma-Aldrich (St. Louis, MO, USA). Phosphate-buffered saline (PBS) was obtained from Solarbio (Beijing, China). MS-222 (tricaine methanesulfonate), at a concentration of 100 mg/L, was purchased from Sigma-Aldrich (St. Louis, MO, USA). The DM and SMX were dissolved in DMSO to obtain a stock solution and then diluted with water to achieve the desired test concentrations.

2.2. Zebrafish Maintenance and Exposure Design

Wild-type AB adult zebrafish (approximately 6 months old) were obtained from the State Key Laboratory of the Chinese Academy of Environmental Sciences. The fish were maintained in a flow-through system at 27 ± 1 °C, 50 ± 10% relative humidity, and a 14 h:10 h light:dark cycle. Continuous aeration in water was employed to maintain dissolved oxygen levels above 7.72 mg/L, while conductivity was kept at 480 μS/cm, with minor fluctuations in these values possible. The fish were acclimatized for 14 days prior to the start of the experiment, during which they were fed freshly hatched brine shrimp (*Artemia nauplii*) twice daily. All experiments were conducted in accordance with the Guidelines for the Keeping and Use of Laboratory Animals of the Chinese Academy of Environmental Sciences. For the acute toxicity test, we conducted the experiment based on the OECD Test Guideline No. 203 for Fish, Acute Toxicity Testing. To avoid potential reproductive or hormonal differences that could influence behavior and neurochemical responses, only male zebrafish were used to assess the toxic effects of DM and SMX in this study. Adult male zebrafish (10 fish per parallel, 6 parallels per concentration group) were exposed to different concentrations of DM (0, 1.0, 2.0, 4.0, 8.0, and 16.0 μg/L) in a clear glass beaker containing 2 L of water for 96 h. Additional groups were exposed to the same concentrations of DM in combination with SMX (30 μg/L) for the same duration. The concentrations of SMX were selected based on environmentally relevant levels [10]. The DM concentrations were chosen based on prior toxicity studies in zebrafish, which demonstrated toxic effects at these concentrations and provided a foundation for evaluating potential ecological risks [6,27].

Each concentration was tested in six replicates to ensure statistical reliability. Control groups were exposed to DMSO at the same concentration as in the experimental groups (0.001% DMSO). Fish were observed for signs of acute toxicity, including mortality, abnormal swimming behavior, and other physical changes (e.g., coloration, fin damage, or erratic movement). At the end of the exposure period (96 h), fish were euthanized using an overdose of MS-222 and thoroughly rinsed and sterilized with PBS, and the following endpoints (mortality, behavioral changes, and immunotoxicity) were assessed. Based on the LC_{50} values, the concentration closest to the LC_{50} was selected for further analysis (4.0 μg/L). This concentration was chosen because it represents a sublethal concentration that ensures sufficient survival rates and it is also suitable for evaluating potential sublethal effects. The control group (0 μg/L) was used to compare and evaluate the baseline endpoints under normal conditions.

2.3. Feeding Behavior Assessment

Food intake was analyzed to assess feeding behavior based on previous studies [20,28]. After 96 h of exposure, a total of 15 zebrafish from each treatment group (2 or 3 in each tank) were randomly collected and placed into three 2 L glass tanks, with five zebrafish per tank and 1 L of clean water in each tank. After a 2 h acclimation period, the zebrafish were fed 30 commercially available food pellets (average diameter = 1 mm; Yee Corp., Weifang,

China). The number of pellets consumed by the zebrafish was recorded at 3, 5, and 10 min post-feeding to assess their feeding behavior. The results were expressed as the ratio of the number of pellets consumed by the zebrafish to the total number of pellets provided.

2.4. Assessment of Neurotransmitter and AChE Levels

The zebrafish brain was carefully dissected, ensuring that the tissue was kept on ice throughout the process to minimize enzymatic degradation. Each brain sample was collected in an ice-cold PBS solution. The brain tissue was homogenized in 10 volumes of ice-cold PBS using a tissue homogenizer or a glass homogenizer. The homogenate was then centrifuged at 4 °C for 15 min at 5000× *g* to remove debris. The supernatant was collected and stored at −80 °C for subsequent analysis.

The total protein concentration in each sample was first measured using a BCA assay, and the data were normalized. The activity of AChE was determined by the enzyme-linked immunosorbent assay with commercial kits (Nanjing Jiancheng Bioengineering Institute, Nanjing, China), following the manufacturer's instructions. The contents of 5-HT and gamma aminobutyric acid (γ-GABA) were measured using the enzyme-linked immunosorbent assay (ELISA) (Shanghai Enzyme-Linked Biotechnology Co., Ltd., Shanghai, China) according to the manufacturer's guidelines. The absorbance (OD value) of AChE and three neurotransmitters was measured by the double-antibody sandwich assay at a wavelength of 450 nm using a enzyme-labeled instrument (BioTek Epoch 2, Agilent Technologies, Inc., Santa Clara, CA, USA). The concentrations of AChE and the three neurotransmitters in the zebrafish samples were calculated using the linear regression equation derived from the standard curve, which was established by plotting the concentrations of the standards against their corresponding OD values. To meet the required sample size for the experiment, we collected the brains from two zebrafish per replicate. Each group included 4 biological replicates (n = 4), and enzyme activity measurements were performed in duplicate.

2.5. Analysis of Oxidative Stress Enzyme Activity

Firstly, zebrafish brain tissue was accurately weighed and added to 9 times the volume of saline in the ratio of weight (g):volume (mL), 1:9, of physiological saline. Mechanical homogenization was performed in an ice-water bath, followed by Eppendorf 5804R centrifugation at 2500 RCF for 10 min(Eppendorf, Hamburg, Germany) and determination of protein content in the supernatant using the kit (Nanjing Jiancheng Bioengineering Institute, Nanjing, China). The activities of superoxide dismutase (SOD), catalase (CAT), glutathione peroxidase (GPX), and glutathione S-transferase (GST) were determined using commercial kits (Nanjing Jiancheng Bioengineering Institute, Nanjing, China) following the manufacturers' instructions. For the assessment of oxidative stress enzyme activity, each treatment group (including the control and experimental groups) consisted of 4 biological replicates (n = 4), and each replicate contained two zebrafish brain tissues. The enzyme activities were measured in duplicate to ensure reliability and consistency of the data.

2.6. Real-Time Quantitative Polymerase Chain Reaction (RT-qPCR) Analysis

Following the 96 h exposure, zebrafish brain tissues were collected for measuring the expression of immune genes to assess the effects of exposure on organisms. Total RNA was extracted from the brain tissues using a commercially available RNA extraction kit (Thermo Fisher Scientific, Waltham, MA, USA) according to the manufacturer's protocol. The extracted RNA was quantified and checked for purity (A260/A280 ratio) by using a Nanodrop ND-2000 spectrophotometer (Thermo Fisher Scientific Inc., Waltham, MA, USA). RNA (1 μg) was reverse transcribed into complementary DNA (cDNA) using a high-capacity cDNA reverse transcription kit (Roche Diagnostics, Basel, Switzerland) following the manufacturer's instructions. RT-qPCR was performed using a Fluorescent

Quantitative PCR Detection System (LineGene 9600, Hangzhou Bioer Technology Co., Ltd., Hangzhou, China) with the FastStart Universal SYBR Green Master (ROX) kit (Roche). Primer sequences for target immune-related genes and housekeeping (RPL13a) genes were referenced to previous literature, and the detailed primer sequences are shown in Table S1. Primers were synthesized by Sangon Biotech (Shanghai, China). The relative expression of the genes was performed for each gene with three technical replicates and three biological replicates (n = 3) and was calculated using the $2^{-\Delta\Delta Ct}$ method.

2.7. Data Statistics and Analysis

The LC_{50} values for deltamethrin and sulfamethoxazole were calculated using Origin 2023 (OriginLab, Northampton, MA, USA). Mortality rates were expressed as percentages for each concentration, and the dose–response data were fitted to a logistic regression model using the Nonlinear Curve Fit tool. The LC_{50} value was derived from the concentration corresponding to 50% mortality based on the logistic equation. The goodness of fit for the logistic model was evaluated using the coefficient of determination (R^2), with a value closer to 1 indicating a good fit. The confidence intervals for the LC_{50} values were calculated, and the precision of the estimates was assessed. Prior to statistical analysis, the normality of the data was evaluated using the Shapiro–Wilk test, which confirmed that the data followed a normal distribution. Therefore, a one-way ANOVA was conducted to analyze the effects on immunogenes, ingestion, neurotransmitters, and oxidative stress enzymes by using GraphPad Prism 9.0 (San Diego, CA, USA). Dunnett's test was used to identify significant differences between the control and treatment groups. Data are presented as mean $\pm$ standard deviation. Differences between groups were considered statistically significant at $p < 0.05$ (*), and highly significant at $p < 0.01$ (**).

3. Results and Discussion

3.1. Acute Toxicity of DM and SMX in Adult Zebrafish

The acute toxicity of DM and SMX was evaluated by calculating the 96 h LC_{50} for adult zebrafish based on statistical mortality, as shown in Figure S1. The LC_{50} for deltamethrin was found to be 4.84 μg/L, indicating significant acute toxicity at relatively low concentrations (Figure 1). The relatively low LC_{50} observed in this study is similar to previous reports highlighting the neurotoxic potential of pyrethroids to aquatic species [29]. Interestingly, the toxicity of the deltamethrin–SMX mixture was lower than that of deltamethrin alone, with the LC_{50} of the mixture being 11.32 μg/L, which is almost twice the LC_{50} of deltamethrin alone. This result suggests a protective or mitigating effect of SMX in reducing the acute toxicity of deltamethrin at the tested concentration. SMX is widely used in aquaculture to control various bacterial diseases due to its ability to inhibit the folate metabolism pathway in bacteria [10]. SMX may alter the pharmacokinetics or bioavailability of deltamethrin, thereby reducing the toxic effects. Deltamethrin exposure in adult zebrafish has been shown to result in similar LC_{50} values as those observed in the current study [30]. These findings suggest that the presence of additional contaminants, such as sulfamethoxazole, may alter deltamethrin toxicity.

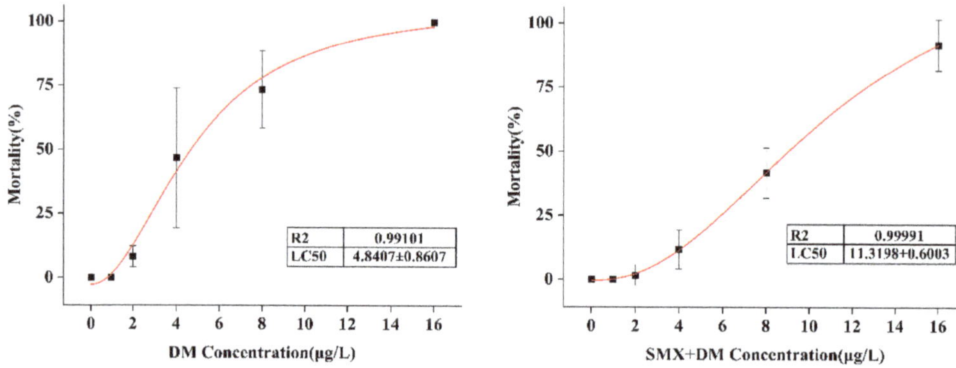

Figure 1. The 96 h LC_{50} results of the acute toxicity test for deltamethrin (DM) and sulfamethoxazole (SMX) on adult zebrafish. The LC_{50} values were determined by plotting the concentration of each chemical against the percentage of mortality, with data points representing the mean mortality rate from multiple replicates. The dose–response curves were fitted using logistic regression, and LC_{50} values were calculated from the concentration corresponding to 50% mortality. Statistical analysis was performed to assess significant differences between the two chemicals.

3.2. Neurotransmitter Alterations in the Zebrafish Brain Following Chemical Exposure

To investigate the neurotoxic effects of DM and its combination with SMX on neurotransmitter levels in adult zebrafish, we quantified γ-GABA, 5-HT, and AChE in the brain 96 h following acute exposure to 4 μg/L of DM and a combination of 4 μg/L of DM and 30 μg/L of SMX (Figure 2). The level of γ-GABA in the brain was significantly altered across the treatment groups. In the control group, the γ-GABA level was 4.94 ± 0.19 μmol/L, while in the DM group and the DM + SMX group, the levels were significantly reduced to 3.91 ± 0.39 μmol/L ($p = 0.002$) and 4.34 ± 0.30 μmol/L ($p = 0.040$), respectively. The significant decrease in γ-GABA levels following DM exposure suggests an impairment of inhibitory neurotransmission, which could be indicative of disruptions in GABAergic signaling pathways [31]. GABA is a crucial inhibitory neurotransmitter in the brain, and alterations in its levels may be associated with neurological disorders, including seizures and anxiety-like behavior [32]. The combined exposure to DM and SMX did not further exacerbate the reduction in γ-GABA levels, suggesting that while SMX may not directly affect GABAergic systems, its presence does not significantly enhance the neurotoxic effects of deltamethrin on GABAergic signaling. This finding indicates that the impact of SMX on neurotransmitter systems may be less pronounced compared to deltamethrin, and that any observed alterations in γ-GABA levels are primarily driven by deltamethrin exposure.

Similarly, the 5-HT levels were found to be lower in both the DM− and DM + SMX-exposed groups compared to the control. The 5-HT concentration was 12.91 ± 0.71 pg/L in the control group, which significantly decreased to 10.88 ± 0.40 pg/L in the DM group ($p < 0.01$) and 11.57 ± 0.90 pg/L ($p < 0.05$) in the DM + SMX group. This suggests potential disturbances in serotonergic signaling, which plays a key role in regulating mood, behavior, and neuroendocrine functions. The reduction in 5-HT, particularly in the DM group, could be linked to the neurotoxic effects of DM, which has been previously reported to affect serotonin receptors and transporters in various organisms [1,33]. The AChE activity was also significantly affected by the chemical exposures. The control group exhibited a level of 11.40 ± 1.03 nmol/L, while the DM group showed a notable decrease to 8.99 ± 0.40 nμmol/L. In the DM + SMX group, the AChE level was partially restored to 11.28 ± 0.94 nmol/L, although this was still lower than the control group. Statistical analysis confirmed a significant reduction in AChE activity in the DM group, but no significant change was observed between the DM + SMX and control groups. AChE activity is essential for the hydrolysis of acetylcholine at synapses, and its inhibition can lead to overstimulation of cholinergic receptors, with subsequent neurological effects [34]. The significant reduction in AChE activity in the DM group indicates potential cholinergic dysfunction, which may contribute to neurobehavioral alterations [27]. Interestingly, the partial restoration of AChE activity in the DM + SMX group suggests that sulfamethoxazole might modulate the impact of cyfluthrin on cholinergic systems, although the exact mechanisms remain unclear.

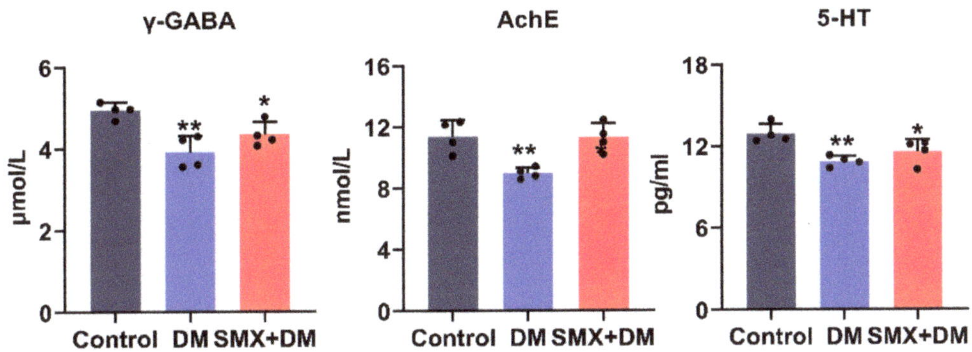

Figure 2. Changes in the levels of major neurotransmitters in the zebrafish brain following 96 h of exposure. Data are shown as mean ± standard deviation (n = 4). Statistical differences between control and exposed groups were assessed using one-way ANOVA, followed by Dunnett's test. As * indicate a significant difference ($p < 0.05$) or ** indicated highly significant ($p < 0.01$) differences.

3.3. Oxidative Stress Response in the Zebrafish Brain After Chemical Exposure

Previous studies have reported oxidative stress effects in zebrafish exposed to deltamethrin (DM), although the specific responses can vary depending on exposure conditions and chemical interactions [27]. To investigate the oxidative stress response in zebrafish brains after acute exposure to 4 μg/L of DM and the combination of 4 μg/L of DM with 30 μg/L of DM + SMX for 96 h, we measured the activities of key antioxidant enzymes, including superoxide SOD, CAT, GPX, and GST (Figure 3).

The results revealed a significant alteration in the levels of these enzymes in the zebrafish brain after chemical exposure. In the control group, the SOD activity was 24.62 ± 1.24 U/mg protein, but it significantly increased to 31.68 ± 1.60 U/mg protein in the DM group ($p < 0.01$) and 26.87 ± 1.48 U/mg protein in the DM + SMX group (Figure 3). The increase in SOD activity in the DM group suggests a compensatory response to oxidative stress induced by cyfluthrin exposure. SOD is the first line of defense against reactive oxygen species (ROS) by catalyzing the conversion of superoxide anions into hydrogen peroxide (H_2O_2), which may indicate an attempt by the zebrafish brain to counteract oxidative damage [35,36]. However, the decrease in SOD activity in the DM + SMX group compared to the DM group suggests that the combination of cyfluthrin and sulfamethoxazole may impair this compensatory mechanism, leading to less efficient ROS neutralization. For CAT, a key enzyme involved in H_2O_2 detoxification [36], the control group exhibited an activity of 19.03 ± 1.04 U/mg protein. After chemical exposure, the levels were significantly increased to 22.87 ± 1.13 U/mg protein in the DM group ($p < 0.05$) and 20.21 ± 2.20 U/mg protein in the DM + SMX group. These reductions reflect a disruption in the detoxification pathway for hydrogen peroxide, a potentially harmful byproduct of cellular metabolism [37]. The observed decrease in CAT activity in both the DM and DM + SMX groups is consistent with the heightened oxidative stress due to chemical exposure, which may overwhelm the cellular antioxidant defenses. Previous studies have reported that DM exposure has been shown to induce oxidative stress in various aquatic organisms, including zebrafish, by disrupting the antioxidant defense systems [38], which supports our findings of significant changes in SOD and CAT activity in zebrafish brain tissues. For instance, a study demonstrated that exposure to deltamethrin led to a significant decrease in SOD and CAT activities in zebrafish liver, suggesting oxidative damage as a result of pesticide toxicity [2]. Similarly, a review study reported that DM exposure in zebrafish embryos resulted in increased oxidative stress markers, including lipid peroxidation, and impaired antioxidant enzyme activities, such as SOD and CAT [27], which aligns with the findings in our study showing an increase in SOD activity in the DM group and a decrease in CAT activity in both the DM and DM + SMX groups.

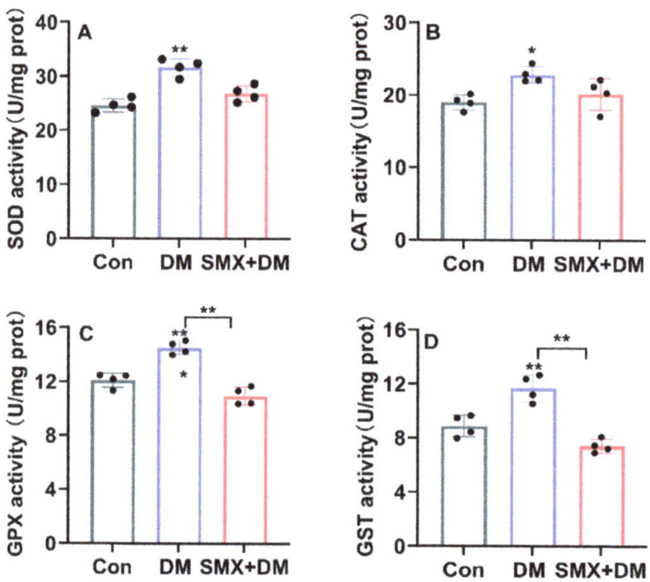

Figure 3. Levels of SOD (**A**), CAT (**B**), GPX (**C**) and GST (**D**) in the zebrafish brain following exposure. Data are expressed as mean ± standard deviation. Statistical analysis was performed using one-way ANOVA, followed by Dunnett's test. Significance levels were denoted as * ($p < 0.05$) for significant differences and ** ($p < 0.01$) for highly significant differences.

The GPX activity in the control group was 12.10 ± 0.52 U/mg protein. In the DM group, GPX activity was significantly reduced to 11.50 ± 0.49 U/mg protein ($p < 0.01$), and it further decreased to 10.95 ± 0.67 U/mg protein in the DM + SMX group. GPX is crucial for reducing lipid peroxides and protecting cell membranes from oxidative damage [37]. The reduction in GPX activity suggests that both cyfluthrin and its combination with sulfamethoxazole induced oxidative damage in zebrafish brain cells, compromising the organism's ability to mitigate lipid peroxidation and oxidative injury [24,39].

GST activity, which plays an important role in detoxification by conjugating reactive electrophiles to glutathione [40], was 8.90 ± 0.82 U/mg protein in the control group. In the DM group, the activity significantly increased to 11.72 ± 0.99 U/mg protein ($p < 0.01$), while in the DM + SMX group, it decreased to 7.44 ± 0.51 U/mg protein. The elevated GST activity in the DM group suggests an enhanced detoxification response to cyfluthrin exposure, possibly due to the increased production of reactive intermediates. However, the decrease in GST activity in the DM + SMX group may indicate an additive or synergistic toxic effect of the two chemicals, which could impair the zebrafish's ability to detoxify harmful compounds, leading to an exacerbation of oxidative stress.

3.4. Changes in Immune-Related Gene Expression in the Zebrafish Brain

We investigated changes in immune-related gene expression in the zebrafish brain following exposure to DM and SMX. Specifically, we assessed the expression of immunoglobulin genes (IgM, IgD, and IgZ) and inflammatory cytokines to understand the immune response alterations induced by these chemicals (Figure 4).

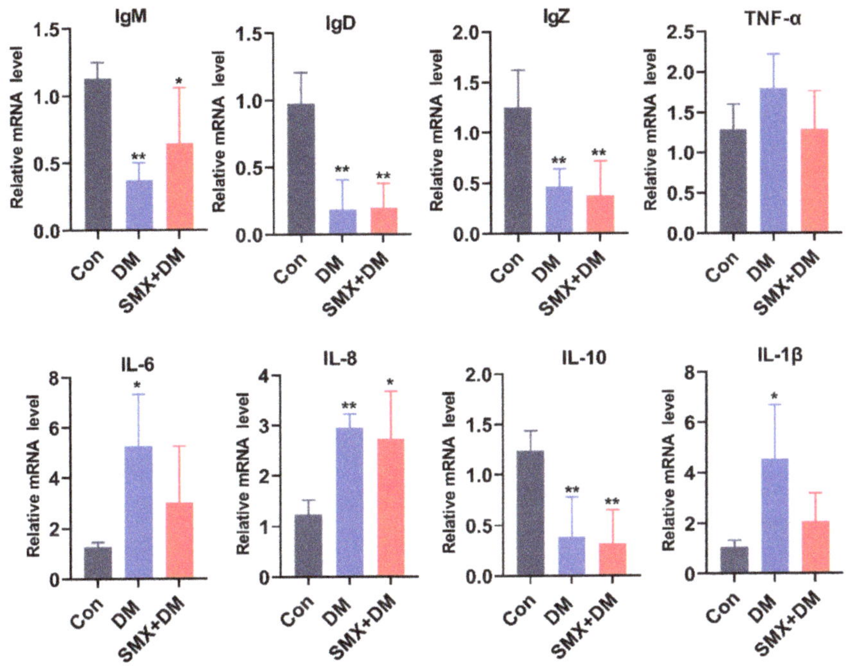

Figure 4. Alterations in the expression of immune-related genes in the zebrafish brain following 96 h of exposure. Gene expression was quantified using quantitative PCR (qPCR) and normalized to the housekeeping gene. Data are presented as fold changes relative to the control group. Statistical significance was determined by one-way ANOVA, followed by Dunnett's test. Statistical significance was indicated by asterisks: * $p < 0.05$ and ** $p < 0.01$.

The expression of IgM, IgD, and IgZ in the zebrafish brain was significantly reduced in both the DM and DM + SMX treatment groups compared to the control group. Specifically, IgM expression decreased from 1.13 ± 0.11-fold in the control group to 0.37 ± 0.13-fold ($p < 0.01$) in the DM group and 0.65 ± 0.41-fold ($p < 0.05$) in the DM + SMX group (Figure 4). Similarly, IgD expression was significantly downregulated in the treatment groups, dropping to 0.19 ± 0.22-fold ($p < 0.01$) in the DM group and 0.20 ± 0.18-fold ($p < 0.01$) in the DM + SMX group, compared to 0.98 ± 0.23-fold in the control group. For IgZ, the control group showed an expression of 1.25 ± 0.36-fold, while both the DM and DM + SMX groups exhibited significant reductions to 0.46 ± 0.17-fold ($p < 0.01$) and 0.37 ± 0.34-fold ($p < 0.01$), respectively. These findings suggest that exposure to DM and SMX suppresses immune-related pathways, particularly those involving immunoglobulins that are crucial for both systemic and mucosal immunity in zebrafish [41], potentially impairing their ability to produce antibodies in response to pathogens.

We also assessed the expression of anti-inflammatory cytokine IL-10. The control group exhibited an expression of 1.24 ± 0.20-fold, whereas the DM and DM + SMX groups had significantly lower expression levels of 0.39 ± 0.39-fold ($p < 0.01$) and 0.32 ± 0.33-fold ($p < 0.01$), respectively. IL-10 is known to regulate inflammatory responses, and its suppression may indicate an impaired ability of the zebrafish to resolve inflammation [42], potentially leading to prolonged immune activation and tissue damage [43,44]. In contrast, IL-8, a pro-inflammatory cytokine, showed an upregulation in both treated groups. The control group exhibited a baseline expression of 1.24 ± 0.28-fold, while the DM and DM + SMX groups had significantly higher levels of 2.95 ± 0.27-fold ($p < 0.01$) and 2.73 ± 0.95-fold ($p < 0.05$), respectively. This suggests that both deltamethrin and sulfamethoxazole induce a pro-inflammatory response in the zebrafish brain, likely triggering the activation of innate immune pathways. IL-8 plays a role in neutrophil recruitment and the overall inflammatory cascade, and its elevation points to an exacerbated inflammatory environment in the brain following chemical exposure [45]. The expression of pro-inflammatory cytokines IL-6 and IL-1β was significantly elevated only in the DM group, with no significant change observed in the DM + SMX group. IL-6 and IL-1β are key players in the initiation of acute inflammation and are known to activate downstream immune signaling pathways [44]. The observed increase in these markers in the DM group suggests that deltamethrin specifically induces an inflammatory response, which may contribute to neuroinflammation and potential neurotoxic effects. Interestingly, the expression of TNF-α did not show significant changes in either the DM or DM + SMX groups compared to the control. Overall, these results highlighted the potential immunotoxic effects of these chemicals, which could impair immune function and contribute to neuroinflammation in aquatic organisms.

3.5. Effects of Deltamethrin and Sulfamethoxazole on Feeding Behavior of Zebrafish

The feeding behavior of zebrafish was assessed by recording the number of pellets consumed at 3, 5, and 10 min after feeding, with results expressed as the ratio of consumed pellets to the total provided (Figure 5). At 3 min, the control group consumed 95.00 ± 4.08% of the pellets, while the DM group showed a significantly lower feeding rate of 70.00 ± 12.25% ($p < 0.01$), which suggested that DM exposure negatively impacts the immediate feeding behavior of zebrafish. However, the DM + SMX group did not significantly alter feeding behavior (85.00 ± 4.08%), indicating a potential mitigating effect of sulfamethoxazole on the feeding impairments caused by deltamethrin. Although feeding rates in the DM and DM + SMX groups slightly decreased at the 5 and 10 min time points compared to the control, no significant differences were observed at these time intervals. The impact on feeding behavior seemed to be short-lived, with no significant differences at later time points.

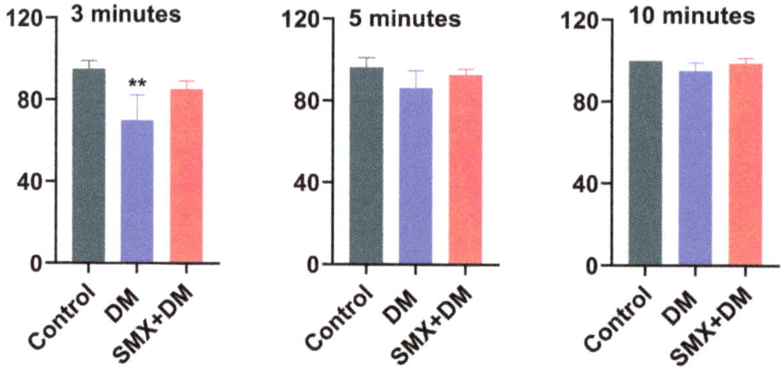

Figure 5. Effects of DM and SMX exposure on the feeding behavior of zebrafish. Statistical differences between the control and exposed groups were assessed using one-way ANOVA, followed by Dunnett's multiple comparison test. ** for $p < 0.01$, indicating highly significant differences.

4. Conclusions

The results of this study demonstrated that DM induced significant acute toxicity, neurotoxic effects, and oxidative stress in zebrafish, while SMX appeared to mitigate some of these effects, particularly in terms of feeding behavior and AChE activity. The combination of DM and SMX did not exacerbate the neurotoxic effects, but partially restored AChE activity. Additionally, oxidative stress responses and immune system alterations suggested that these chemicals can impair the zebrafish's ability to cope with environmental stressors. The observed decrease in immune-related gene expression and increase in pro-inflammatory cytokines indicated potential immunotoxicity, which may contribute to neuroinflammation. These findings underscore the importance of considering chemical interactions in environmental risk assessments and highlight the need for further research to explore the underlying mechanisms of toxicity and their long-term impacts on aquatic ecosystems.

Supplementary Materials: The following supporting information can be downloaded at: https://www.mdpi.com/article/10.3390/toxics13020128/s1. Figure S1: Mortality of zebrafish after 96 h of acute exposure to different concentrations of DM and SMX. Table S1: Primers used for real-time quantative PCR assays [28,46].

Author Contributions: Conceptualization, Y.L. and F.L.; methodology, Y.L.; software, Y.L.; validation, Y.L. and C.W.; formal analysis, C.W.; investigation, Y.L.; data curation, Y.L.; writing—original draft preparation, Y.L.; writing—review and editing, C.W.; visualization, Y.L.; supervision, F.L.; funding acquisition, C.W. All authors have read and agreed to the published version of the manuscript.

Funding: This research was funded by the National Natural Science Foundation of China, grant number 42377274, and the National Key Research and Development Program of China, grant number 2021YFC3200801.

Institutional Review Board Statement: All operations were supervised by certified professionals accredited by the Chinese Association for Laboratory Animal Sciences (CALAS), ensuring compliance with animal welfare measures (e.g., anesthesia, euthanasia) as stipulated in the Technical Specifications for the Humane Euthanasia of Laboratory Animals (GB/T 39760-2021).

Informed Consent Statement: Not applicable.

Data Availability Statement: The datasets used and/or analyzed during the current study are available from the corresponding author upon reasonable request.

Conflicts of Interest: The authors declare that they have no known competing financial interests or personal relationships that could have appeared to influence the work reported in this paper.

References

1. Pitzer, E.M.; Williams, M.T.; Vorhees, C.V. Effects of pyrethroids on brain development and behavior: Deltamethrin. *Neurotoxicol. Teratol.* **2021**, *87*, 106983. [CrossRef]
2. Kumar, A.; Jasrotia, S.; Dutta, J.; Kyzas, G.Z. Pyrethroids toxicity in vertebrates and invertebrates and amelioration by bioactive compounds: A review. *Pestic. Biochem. Physiol.* **2023**, *196*, 105615. [CrossRef] [PubMed]
3. Wolansky, M.J.; Harrill, J.A. Neurobehavioral toxicology of pyrethroid insecticides in adult animals: A critical review. *Neurotoxicol. Teratol.* **2008**, *30*, 55–78. [CrossRef] [PubMed]
4. Feo, M.L.; Ginebreda, A.; Eljarrat, E.; Barceló, D. Presence of pyrethroid pesticides in water and sediments of Ebro River Delta. *J. Hydrol.* **2010**, *393*, 156–162. [CrossRef]
5. Lao, W.; Tsukada, D.; Greenstein, D.J.; Bay, S.M.; Maruya, K.A. Analysis, occurrence, and toxic potential of pyrethroids, and fipronil in sediments from an urban estuary. *Environ. Toxicol. Chem.* **2010**, *29*, 843–851. [CrossRef] [PubMed]
6. Ortiz-Pérez, M.D.; Torres-Dosal, A.; Batres, L.E.; López-Guzmán, O.D.; Grimaldo, M.; Carranza, C.; Pérez-Maldonado, I.N.; Martínez, F.; Pérez-Urizar, J.; Díaz-Barriga, F. Environmental health assessment of deltamethrin in a malarious area of Mexico: Environmental persistence, toxicokinetics, and genotoxicity in exposed children. *Environ. Health Perspect.* **2005**, *113*, 782–786. [CrossRef]

7. Fang, W.; Peng, Y.; Muir, D.; Lin, J.; Zhang, X. A critical review of synthetic chemicals in surface waters of the US, the EU and China. *Environ. Int.* **2019**, *131*, 104994. [CrossRef]
8. Peris, A.; Barbieri, M.V.; Postigo, C.; Rambla-Alegre, M.; López de Alda, M.; Eljarrat, E. Pesticides in sediments of the Ebro River Delta cultivated area (NE Spain): Occurrence and risk assessment for aquatic organisms. *Environ. Pollut.* **2022**, *305*, 119239. [CrossRef] [PubMed]
9. Oliveira, J.M.; Lima, G.D.A.; Destro, A.L.F.; Condessa, S.; Zuanon, J.A.S.; Freitas, M.B.; Oliveira, L.L. Short-term intake of deltamethrin-contaminated fruit, even at low concentrations, induces testicular damage in fruit-eating bats (*Artibeus lituratus*). *Chemosphere* **2021**, *278*, 130423. [CrossRef] [PubMed]
10. Zhang, R.; Pei, J.; Zhang, R.; Wang, S.; Zeng, W.; Huang, D.; Wang, Y.; Zhang, Y.; Wang, Y.; Yu, K. Occurrence and distribution of antibiotics in mariculture farms, estuaries and the coast of the Beibu Gulf, China: Bioconcentration and diet safety of seafood. *Ecotoxicol. Environ. Saf.* **2018**, *154*, 27–35. [CrossRef]
11. Radke, M.; Lauwigi, C.; Heinkele, G.; Mürdter, T.E.; Letzel, M. Fate of the antibiotic sulfamethoxazole and its two major human metabolites in a water sediment test. *Environ. Sci. Technol.* **2009**, *43*, 3135–3141. [CrossRef]
12. Mo, W.Y.; Chen, Z.; Leung, H.M.; Leung, A.O. Application of veterinary antibiotics in China's aquaculture industry and their potential human health risks. *Environ. Sci. Pollut. Res. Int.* **2017**, *24*, 8978–8989. [CrossRef]
13. Kemper, N. Veterinary antibiotics in the aquatic and terrestrial environment. *Ecol. Indic.* **2008**, *8*, 1–13. [CrossRef]
14. Straub, J.O. Aquatic environmental risk assessment for human use of the old antibiotic sulfamethoxazole in Europe. *Environ. Toxicol. Chem.* **2016**, *35*, 767–779. [CrossRef] [PubMed]
15. Jijie, R.; Solcan, G.; Nicoara, M.; Micu, D.; Strungaru, S.A. Antagonistic effects in zebrafish (*Danio rerio*) behavior and oxidative stress induced by toxic metals and deltamethrin acute exposure. *Sci. Total Environ.* **2020**, *698*, 134299. [CrossRef] [PubMed]
16. Trush, M.A.; Thompson, D.C. Enhancement of Chemical Activation via Radical-Dependent Mechanisms: An Emerging Concept in Chemical-Chemical Interactions. In *Oxygen Radicals in Biology and Medicine*; Simic, M.G., Taylor, K.A., Ward, J.F., von Sonntag, C., Eds.; Springer: Boston, MA, USA, 1988; pp. 739–744. [CrossRef]
17. Desalegn, A.; Bopp, S.; Asturiol, D.; Lamon, L.; Worth, A.; Paini, A. Role of Physiologically Based Kinetic modelling in addressing environmental chemical mixtures—A review. *Comput. Toxicol.* **2019**, *10*, 158–168. [CrossRef] [PubMed]
18. Sipes, N.S.; Padilla, S.; Knudsen, T.B. Zebrafish: As an integrative model for twenty-first century toxicity testing. *Birth Defects Res. Part C Embryo Today Rev.* **2011**, *93*, 256–267. [CrossRef] [PubMed]
19. He, J.H.; Gao, J.M.; Huang, C.J.; Li, C.Q. Zebrafish models for assessing developmental and reproductive toxicity. *Neurotoxicol. Teratol.* **2014**, *42*, 35–42. [CrossRef]
20. Blanco, A.M.; Bertucci, J.I.; Hatef, A.; Unniappan, S. Feeding and food availability modulate brain-derived neurotrophic factor, an orexigen with metabolic roles in zebrafish. *Sci. Rep.* **2020**, *10*, 10727. [CrossRef] [PubMed]
21. Rico, E.P.; Rosemberg, D.B.; Seibt, K.J.; Capiotti, K.M.; Da Silva, R.S.; Bonan, C.D. Zebrafish neurotransmitter systems as potential pharmacological and toxicological targets. *Neurotoxicol. Teratol.* **2011**, *33*, 608–617. [CrossRef]
22. Parng, C.; Roy, N.M.; Ton, C.; Lin, Y.; McGrath, P. Neurotoxicity assessment using zebrafish. *J. Pharmacol. Toxicol. Methods* **2007**, *55*, 103–112. [CrossRef] [PubMed]
23. Bally-Cuif, L.; Vernier, P. 2-Organization and physiology of the zebrafish nervous system. In *Fish Physiology*; Perry, S.F., Ekker, M., Farrell, A.P., Brauner, C.J., Eds.; Academic Press: Cambridge, MA, USA, 2010; Volume 29, pp. 25–80.
24. Balmus, I.M.; Ciobica, A.; Antioch, I.; Dobrin, R.; Timofte, D. Oxidative Stress Implications in the Affective Disorders: Main Biomarkers, Animal Models Relevance, Genetic Perspectives, and Antioxidant Approaches. *Oxidative Med. Cell. Longev.* **2016**, *2016*, 3975101. [CrossRef]
25. Huh, J.R.; Veiga-Fernandes, H. Neuroimmune circuits in inter-organ communication. *Nat. Rev. Immunol.* **2020**, *20*, 217–228. [CrossRef]
26. Dantzer, R.; O'Connor, J.C.; Freund, G.G.; Johnson, R.W.; Kelley, K.W. From inflammation to sickness and depression: When the immune system subjugates the brain. Nature reviews. *Neuroscience* **2008**, *9*, 46–56. [CrossRef]
27. Parlak, V. Evaluation of apoptosis, oxidative stress responses, AChE activity and body malformations in zebrafish (*Danio rerio*) embryos exposed to deltamethrin. *Chemosphere* **2018**, *207*, 397–403. [CrossRef]
28. Liu, Y.; Lin, S.; Wang, C.; Li, T.; Zheng, G.; Sun, W.; An, L.; Bai, Y.; Wu, F. Sex-Specific Effects of Environmental Exposure to the Antimicrobial Agents Benzalkonium Chloride and Triclosan on the Gut Microbiota and Health of Zebrafish (*Danio rerio*). *Environ. Sci. Technol.* **2024**, *58*, 15450–15462. [CrossRef] [PubMed]
29. Ren, Q.; Zhang, T.; Li, S.; Ren, Z.; Yang, M.; Pan, H.; Xu, S.; Qi, L.; Chon, T.S. Integrative Characterization of Toxic Response of Zebra Fish (*Danio rerio*) to Deltamethrin Based on AChE Activity and Behavior Strength. *BioMed Res. Int.* **2016**, *2016*, 7309184. [CrossRef] [PubMed]
30. Görge, G.; Nagel, R. Toxicity of lindane, atrazine, and deltamethrin to early life stages of zebrafish (*Brachydanio rerio*). *Ecotoxicol. Environ. Saf.* **1990**, *20*, 246–255. [CrossRef] [PubMed]

31. Lummis, S.C.R.; Chow, S.C.; Holan, G.; Johnston, G.A.R. γ-Aminobutyric Acid Receptor Ionophore Complexes: Differential Effects of Deltamethrin, Dichlorodiphenyltrichloroethane, and Some Novel Insecticides in a Rat Brain Membrane Preparation. *J. Neurochem.* **1987**, *48*, 689–694. [CrossRef]
32. Alasmari, F. Caffeine induces neurobehavioral effects through modulating neurotransmitters. *Saudi Pharm. J. SPJ Off. Publ. Saudi Pharm. Soc.* **2020**, *28*, 445–451. [CrossRef]
33. Lee, J.G.; Cho, H.J.; Jeong, Y.M.; Lee, J.S. Genetic Approaches Using Zebrafish to Study the Microbiota-Gut-Brain Axis in Neurological Disorders. *Cells* **2021**, *10*, 566. [CrossRef]
34. Muthulakshmi, S.; Maharajan, K.; Habibi, H.R.; Kadirvelu, K.; Venkataramana, M. Zearalenone induced embryo and neurotoxicity in zebrafish model (*Danio rerio*): Role of oxidative stress revealed by a multi biomarker study. *Chemosphere* **2018**, *198*, 111–121. [CrossRef]
35. Valavanidis, A.; Vlahogianni, T.; Dassenakis, M.; Scoullos, M. Molecular biomarkers of oxidative stress in aquatic organisms in relation to toxic environmental pollutants. *Ecotoxicol. Environ. Saf.* **2006**, *64*, 178–189. [CrossRef]
36. Deisseroth, A.; Dounce, A.L. Catalase: Physical and chemical properties, mechanism of catalysis, and physiological role. *Physiol. Rev.* **1970**, *50*, 319–375. [CrossRef]
37. Livingstone, D.R. Contaminant-stimulated reactive oxygen species production and oxidative damage in aquatic organisms. *Mar. Pollut. Bull.* **2001**, *42*, 656–666. [CrossRef]
38. Yang, C.; Lim, W.; Song, G. Mediation of oxidative stress toxicity induced by pyrethroid pesticides in fish. Comparative biochemistry and physiology. *Toxicol. Pharmacol. CBP* **2020**, *234*, 108758. [CrossRef]
39. Kurutas, E.B. The importance of antioxidants which play the role in cellular response against oxidative/nitrosative stress: Current state. *Nutr. J.* **2016**, *15*, 71. [CrossRef]
40. Li, S.; Tan, H.Y.; Wang, N.; Zhang, Z.J.; Lao, L.; Wong, C.W.; Feng, Y. The Role of Oxidative Stress and Antioxidants in Liver Diseases. *Int. J. Mol. Sci.* **2015**, *16*, 26087–26124. [CrossRef]
41. Mashoof, S.; Criscitiello, M.F. Fish Immunoglobulins. *Biology* **2016**, *5*, 45. [CrossRef]
42. Ouyang, W.; Rutz, S.; Crellin, N.K.; Valdez, P.A.; Hymowitz, S.G. Regulation and functions of the IL-10 family of cytokines in inflammation and disease. *Annu. Rev. Immunol.* **2011**, *29*, 71–109. [CrossRef]
43. Iyer, S.S.; Cheng, G. Role of interleukin 10 transcriptional regulation in inflammation and autoimmune disease. *Crit. Rev. Immunol.* **2012**, *32*, 23–63. [CrossRef]
44. Chen, L.; Deng, H.; Cui, H.; Fang, J.; Zuo, Z.; Deng, J.; Li, Y.; Wang, X.; Zhao, L. Inflammatory responses and inflammation-associated diseases in organs. *Oncotarget* **2018**, *9*, 7204–7218. [CrossRef]
45. Goldstein, B.I.; Kemp, D.E.; Soczynska, J.K.; McIntyre, R.S. Inflammation and the phenomenology, pathophysiology, comorbidity, and treatment of bipolar disorder: A systematic review of the literature. *J. Clin. Psychiatry* **2009**, *70*, 1078–1090. [CrossRef]
46. Medina-Gali, R.M.; Ortega-Villaizan, M.D.M.; Mercado, L.; Novoa, B.; Coll, J.; Perez, L. Beta-glucan enhances the response to SVCV infection in zebrafish. *Dev. Comp. Immunol.* **2018**, *84*, 307–314. [CrossRef]

toxics

MDPI

Article

Influencing Factors of Phosphorus Mobility and Retention in the Sediment of Three Typical Plateau Lakes

Xue Wu [1,2,3], Yancai Wang [2], Lixin Jiao [3,*], Jia He [2,*], Hongbin Zhou [2] and Zhengzheng Hao [4]

1 Faculty of Environmental Science and Engineering, Kunming University of Science and Technology, Kunming 650500, China; stellawu7@126.com
2 Kunming Institute of Eco-Environmental Sciences, Kunming 650032, China; diamond_wyc@163.com (Y.W.); zhouh.b@163.com (H.Z.)
3 State Key Laboratory of Environmental Criteria and Risk Assessment, Chinese Research Academy of Environmental Sciences, Beijing 100012, China
4 Yunnan Dianwei Environmental Protection Technology Co., Ltd., Kunming 650031, China; haozhengzheng2024@163.com
* Correspondence: jiaolx@craes.org.cn (L.J.); dcszxb@163.com (J.H.)

Abstract: The mechanisms driving changes in the stability of phosphorus (P) in sediments under lake ecosystem degradation remain poorly understood. This study investigated the P-binding forms in sediments from three plateau lakes with different trophic states in Yunnan Province, China, aiming to elucidate the responses of sediment P compositions to human activities, lake trophic status, and dissolved organic matter (DOM) characteristics. The results showed that human activity directly contributed to sediment P retention. The trophic type of lake exerted a discernible effect on P mobility in the sediments, as eutrophic algae-type lakes had a higher content of sediment mobile-P. Moreover, the sediment DOM promoted the adsorption of BD-P and NH_4Cl-P. Generally, exogenous pollution caused by human activity leads to lake eutrophication and a decline in lake ecosystem stability. This variation was largely influenced by water depth. A decrease in lake ecosystem stability leads to increased P mobility in sediments, which increases the risk of endogenous pollution. The DOM plays an important role in the mobility of sediment P. These insights offer a novel perspective for understanding how lake ecosystem characteristics are related to endogenous P loads in lakes.

Keywords: lakes; sediments; phosphorus; retention; mobility

1. Introduction

Phosphorus (P) is a crucial nutrient for aquatic vegetation growth; however, its excessive accumulation in water bodies leads to eutrophication, characterized by cyanobacterial blooms and the production of phycotoxins, posing severe ecotoxicological risks to aquatic ecosystems [1–3]. Lake sediments act as major reservoirs of P. Even when external P inputs are controlled, the release of internal P from sediments remains a key driver of lake eutrophication [4–6]. Sedimentary P exists in two main forms: mobile-P and immobile-P. Mobile-P includes weakly adsorbed phosphorus (NH_4Cl-P), organic phosphorus (NaOH-nrP), and iron-bound phosphorus (BD-P), while immobile-P includes aluminum-bound phosphorus (NaOH-rP), calcium-bound phosphorus (HCl-P), and residual phosphorus (Res-P), each exhibiting distinct bioavailability and environmental behavior [7–10]. Among these, mobile-P is highly bioavailable and can readily support algal growth [11–13].

Dissolved organic matter (DOM), as the most reactive and bioavailable component of sedimentary organic matter (SOM), plays a pivotal role in lake eutrophication by influencing the geochemical cycling of P [14,15]. DOM includes various functional groups, such as carboxyl, hydroxyl, amino, and aromatic groups, that interact with P, as well as metal ions, to create complexes that affect the environmental behavior and bioavailability of P [16,17]. The interactions between DOM and P are variable and complex. For instance, the carboxyl groups in DOM can act as organic anions to partially replace phosphate adsorbed by the soil through competitive adsorption. Alternatively, soluble ternary complexes such as DOM-Fe-PO_4 and DOM-Al-PO_4 can be formed through complexation and dissolution, thereby enhancing P bioavailability [18,19]. Kurek et al. [20] highlighted the pivotal role of DOM in seasonal organophosphorus transformation. Humic-like components in DOM are recognized as key factors governing P adsorption and release [21]. Recent research has indicated that the distribution of the fluorescent components in DOM is closely associated with the adsorption characteristics of P in the sediments of Erhai Lake [22]. High levels of protein-like components can reduce P adsorption in sediments by decreasing the goethite content [22,23]. Furthermore, in Taihu Lake, DOM resulting from algal blooms under eutrophic conditions enhanced the reduction and dissolution of Fe and Mn oxides, and this process inevitably influenced the forms of Fe- and Mn-bound P in sediments, as well as the release of P [24].

Many researchers have studied P forms, DOM characteristics, and the binding mechanisms between P and DOM in lake sediments [25–27]. Zhao et al. [28] investigated how the composition and structural characteristics of organic matter affect the phosphorus forms in the sediments of Yangzonghai Lake. Chen et al. [29] investigated the effects of climate change and human activity on the P of the sediments in Dianchi Lake. Ni et al. [30] studied the effects of DOM on the degradation and release of diester P in the sediments of the low-nutrient Erhai Lake. However, many of these studies have been limited to individual lakes, with few reports on the binding forms of sediment P in lakes with different anthropogenic intensities and trophic states and their correlation with DOM. The origins of P and DOM in lake sediments encompass endogenous and exogenous sources, with external human activities and the hydrological, water quality, and ecological conditions of the lake, which play formidable roles in shaping the characteristics of DOM and its binding behavior with P [31,32]. This study investigated the forms of phosphorus in sediments from three plateau lakes in Yunnan Province, China: Dianchi Lake, Erhai Lake, and Yangzonghai Lake, each representing a different trophic state. It presents a comparative analysis of the basic characteristics of these lakes, the occurrence of sediment P forms, and the fluorescence characteristics of sediment DOM. The main objectives of this paper were to examine how DOM and human activities influence phosphorus retention in sediments and to explore how lake nutrient levels and ecological characteristics impact the mobility of phosphorus in lake sediments.

2. Materials and Methods

2.1. Overview of the Research Areas

Yunnan Province in China is home to Dianchi Lake, Erhai Lake, and Yangzonghai Lake, all of which display distinct variations in basin size, water depth, level of human activity, and trophic state.

Situated in the southern part of Kunming main city area, Dianchi Lake (N24°40′–25°02′, E102°36′–102°47′) is recognized as the sixth-largest freshwater lake in China. Dianchi Lake is separated into Caohai Lake and Waihai Lake by a dam, with a total basin area of 2920 km^2 and a water area of 297.9 km^2 [33]. Dianchi Lake is classified as an urban lake with a population of 5,453,500. The water quality of Dianchi Lake has deteriorated due to social and economic development since the late 1970s, leading to severe eutrophication and consistently low water quality below Grade V. However, after years of extensive treatment, the water quality was upgraded to Grade IV by 2018. However, it is still characterized as a typical eutrophic algal lake with a submerged vegetation coverage rate of approximately 3% [34]. Based on the 2018 investigation data on sediments in Dianchi Lake and the findings of Zhang et al. [35], the surface sediments of the lake are primarily black in color, with a foul odor, and the particle size is predominantly within the <250 μm range, with silt (4 to <63 μm) accounting for the largest proportion, averaging 60%, followed by clay (<4 μm), which constitutes an average of 19%.

Situated to the north of Dali City, Erhai Lake (N25°35′–25°58′, E100°5′–100°17′) is the seventh-largest freshwater lake in China. The lake covers a basin measuring 2565 km^2, with a water surface area of 249.8 km^2. This lake houses a basin population of 854,800 [36]. Since the 1990s, rapid economic development has caused a surge in the pollution load of the lake, leading to consistent elevation of nutrients in the lake. Consequently, the trophic status has gradually shifted from oligotrophic to mesotrophic. Currently, the water quality of Erhai Lake is categorized as Grades II–III, and the lake is in a transitional phase from mesotrophic to eutrophic. As a grass–algae mixed lake, it houses relatively abundant aquatic vegetation, with a vegetation coverage of approximately 10%. Although large-scale cyanobacterial blooms occurred in 1996, 2003, and 2013, the water quality of this lake has remained satisfactory [37]. The sediments of Erhai Lake appear brown, exhibiting a foul odor. The sediment primarily comprises clay, silt, fine sand, and coarse sand. The clay particles range from 0.57 to 1.1 μm, with a percentage content varying between 2.6% and 12.2%, and the fine silt particles range from 1.5–6.4 μm, with a percentage content ranging from 7.5% to 89.8% [38].

Situated in the southeastern part of Kunming, Yangzonghai Lake (N24°51′–24°58′, E102°58′–103°01′) spans 286 km^2 in basin area and 31.3 km^2 in water area. The lake is a typical deep-water lake with a population of 61,400. Yangzonghai Lake was considered oligotrophic before the 1990s; however, the introduction of cage fish culture in the late 1990s accelerated the decline in water quality. Moreover, the period from 1996 to 1998 witnessed multiple occurrences of cyanobacterial blooms. In 2008, the lake experienced serious As contamination stemming from industrial operations. The water depth of the lake makes the aquatic vegetation community relatively simple, mainly distributed in the shallow water area of the lake bank. Presently, this lake is categorized as a mesotrophic algae lake [39]. The surface sediments of Yangzonghai Lake are brown and have a slight odor. They consist primarily of clay, fine silt, and sand. The clay particles range from 0.57 to 1.1 μm, accounting for 2.6% to 12.2% of the sediment, and the fine silt particles range from 1.5 to 6.4 μm, comprising 7.5% to 89.8% of the total sediment [38].

More information on the basic characteristics of the three lakes can be found in Section S1 of the Supplementary Material.

2.2. Sampling Site Layout

Ten sediment sampling sites were set up in the Dianchi, Erhai, and Yangzonghai lakes. Among them, four monitoring sites were set up in Dianchi Lake, located in Caohai Lake, as well as the northern, middle, and southern regions of Waihai Lake, which were numbered C1, W1, W2, and W3, respectively. Similarly, three monitoring sites were set up in Erhai Lake, located in the northern, central, and southern parts of Erhai Lake, denoted as E1, E2, and E3, respectively. Three monitoring sites were set up in Yangzonghai Lake, located in the north, center, and south of the lake and labelled Y1, Y2, and Y3, respectively. The specific sampling site positions are illustrated in Figure 1.

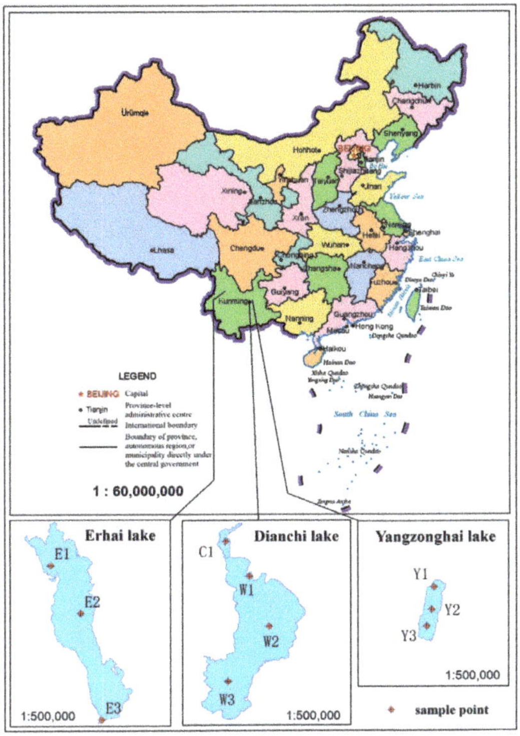

Figure 1. Layout of sediment sampling sites.

2.3. Sampling and Analysis

2.3.1. Sediment Sampling

In March 2021, surficial sediment samples from 10 monitoring sites were collected using a mud bucket at a depth of 0–10 cm. The samples were carefully sealed in airtight plastic bags and placed in refrigerated storage before being transported to the laboratory for natural air drying. After air drying, surplus organic and inorganic residues were removed from the sediment samples, which were subsequently pulverized into a fine powder using a sieve for future analysis [25].

2.3.2. Analysis of SOM and DOM

The SOM content of 10 sediment samples collected from Dianchi Lake, Erhai Lake, and Yangzonghai Lake was determined using the weight difference method [40]. After adjusting the moisture content, 40 g of the treated sediment was combined with deionized water in a 1:5 ratio. The blend was agitated at 300 rpm for a full day, followed by centrifugation at 4000 rpm for 10 min. The supernatant was filtered through a 0.45 μm glass fiber filter (pre-fired at 450 °C and maintained at a constant temperature for 5 h). The filtered solution, referred to as DOM solution for the experiment, was stored in the dark at 4 °C [41].

Three-dimensional fluorescence spectra of the DOM samples were analyzed. The excitation wavelength range was 200–450 nm, with a scanning interval of 5 nm. With a similar scanning interval, the emission wavelength range was set to 250–600 nm. The slits of the excitation and emission units were adjusted to 10 nm, the photomultiplier tube was operated at 400 V, and the scanning speed was settled to 2400 nm/min.

Fluorescence data were analyzed using the PARAFAC method to identify the DOM fluorescence components and their intensities. The data were imported into 3D specT-ALYZE 1.5 for noise removal and scattering correction. A three-dimensional projection or contour map was created based on the excitation and emission wavelengths and intensities. The specific excitation and emission spectra of each component (C1, C2, C3, and C4) were compared with the OpenFluor database to identify the corresponding substances.

2.3.3. Determination and Analysis of P

The sediment sample weighing 0.2 g underwent combustion at 450 °C for 3 h and was then transferred into a 50 mL centrifugal tube for centrifugation. The total phosphorus (TP) was quantified using standard methods of alkaline K persulfate digestion and ultraviolet spectrophotometry [42]. The sediment P fractions were extracted using the continuous extraction method [9,43]. In the extraction process, 1 g of sediment samples were first weighed. Then, NH_4Cl-P, BD-P, NaOH-rP, HCl-P, and Res-P were extracted with NH_4Cl (1 M), $Na_2S_2O_4/NaHCO_3$, NaOH (0.1 M), HCl (0.5 M), and NaOH (1 M) reagents, respectively. After extraction, the above extraction solution was centrifugated for 10 min at 4000 r/min and filtered through a 0.45 um filter membrane. The content of phosphorus in the extraction solution was determined using molybdenum–antimonic spectrophotometry. NaOH-nrP was obtained by subtracting the above inorganic phosphorus from the TP.

2.3.4. Water Quality Analysis

The water quality monitoring parameters in this study consisted of chemical oxygen demand (COD), TP, ammonia nitrogen (NH_3-N), total nitrogen (TN), chlorophyll a (Chla), and the trophic state index. The preservation and monitoring techniques used for the samples were based on the guidelines provided in the 4th Edition of Monitoring and Analysis Methods for Water and Wastewater [44].

2.3.5. Data Analysis and Quality Assurance

We analyzed the correlations between sediment phosphorus forms and both sediment DOM characteristics and human activity indicators in the watershed. The Pearson correlation coefficient was used to examine the relationships between these variables in the three lakes. The sediment indicators included TP content, phosphorus forms, DOM content, SOM mass fraction, and fluorescence intensity of the four components. The human activity indicators included GDP per unit area, population density, urbanization rate (urban population/total population), and the proportion of urban area (urban land area/total land area). A Pearson correlation analysis was conducted, and correlation matrix heatmaps were generated. Significant correlations were considered with p-values < 0.05.

In this paper, graphs were created using ArcGIS 10.8 and Origin 2021. To ensure data reliability and accuracy, 3 samples were collected at each site, mixed for representativeness, and promptly transported to the lab. The samples were sieved through a 100-mesh sieve for uniformity, and repeated experiments and blank tests were conducted. All reagents used were of analytical grade or higher, meeting industry standards and being used before their expiration dates.

3. Results

3.1. Basic Features of the Three Typical Lakes

Among the three typical plateau lakes investigated in this study, Dianchi Lake exhibited moderate pollution, with water quality classified as Class IV, while Erhai Lake and Yangzonghai Lake had relatively good water quality, both classified as Class II. The chlorophyll a (Chla) concentrations in Dianchi Lake were 10.66 and 11.88 times greater than those in Erhai Lake and Yangzonghai Lake, respectively. Similarly, the trophic level index was 1.58 times higher than in Erhai Lake and 1.51 times higher than in Yangzonghai Lake. The TP concentration of Dianchi Lake water was about 0.06 mg/L, and those of Erhai Lake and Yangzonghai Lake were both about 0.02 mg/L.

Analysis of the sediments from the three lakes indicated that the average contents of SOM and DOM in Dianchi Lake were 19.49% and 1640.63 mg/kg, respectively, which were the highest among the three lakes. Both SOM and DOM peaked in the Caohai region, significantly surpassing other areas. In contrast, Erhai Lake and Yangzonghai Lake had only 24% and 32% of the SOM content of Dianchi Lake, respectively, with their DOM contents at approximately 62% and 81% of Dianchi Lake's levels. Spatial variation in SOM and DOM distribution was relatively low in Erhai and Yangzonghai Lakes. The sediment DOM of Erhai Lake was mainly concentrated in the north and south and was relatively low in the middle area. The sediment DOM of Yangzonghai Lake was the highest in the middle, second-highest in the south, and lowest in the north.

Dianchi Lake also exhibited significantly higher porewater SRP concentrations compared to Erhai Lake and Yangzonghai Lake: 7.91 and 6.06 times higher, respectively. The SRP concentrations were highest in the Caohai region and the northern part of Waihai, while the middle and southern parts of Waihai showed relatively low concentrations. In Erhai Lake, the SRP levels were higher in the middle and southern regions compared to the northern part, where sediment acted in an "adsorption" state, with SRP concentrations lower than those in the overlying water. In Yangzonghai Lake, the SRP concentrations were highest in the middle, followed by the north and lowest in the south.

Detailed information can be found in Section S2 of the Supplementary Material.

3.2. Characteristics of P-Binding Forms in Sediments

The average sediment TP contents in the Dianchi, Erhai, and Yangzonghai lakes were 2395.24, 1017.49, and 896.64 mg/kg, respectively. Dianchi Lake has a relatively high TP content, approximately twice that of Yangzonghai and Erhai Lakes. Different P-binding forms in the sediments demonstrate variations in P mobility and bioavailability [45]. Figure 2 shows the contents of different forms of P in the sediments of the Dianchi, Erhai, and Yangzonghai lakes. In Caohai Lake, the predominant form of P was NaOH-rP, which accounted for 34.72% of the total P content. Conversely, HCl-P was the featured P form in the Waihai and Erhai lakes, accounting for 33.10% and 28.70%, respectively, on average. In particular, the HCl-P in Waihai Lake was mainly distributed in the central and southern regions, whereas the other P forms were principally distributed in the northern region. HCl-P in Erhai Lake prevails extensively in the northern region, whereas other forms are concentrated in the central region. Res-P emerged as the leading form of P in Yangzonghai Lake, accounting for 25.99% of the P content. Regardless of form, the P content gradually decreased from north to south in Yangzonghai Lake. NH_4Cl-P is the least concentrated form of P in lake sediments, comprising <1% of the TP. Being highly biologically active, NH_4Cl-P is readily released into water, absorbed, and used by aquatic plants, resulting in a low content in the sediment [46].

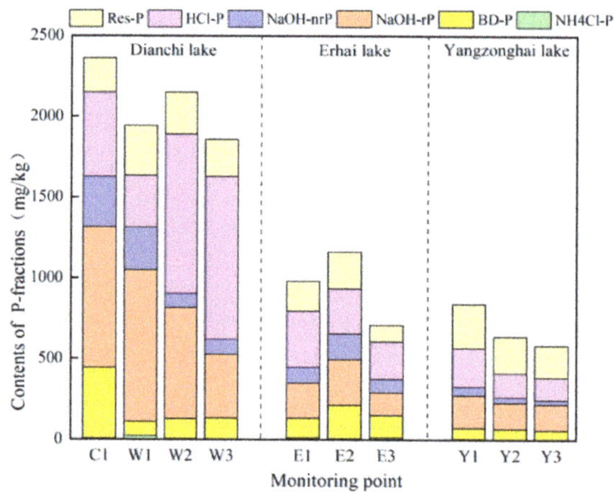

Figure 2. P form contents in the sediments of various monitoring sites in the lakes.

3.3. DOM Fluorescence Characteristics of Sediments

The DOM contents in the sediments of Dianchi Lake, Erhai Lake, and Yangzonghai Lake were analyzed based on their three-dimensional fluorescence spectrum. Details of the fluorescent components of sediment DOM in the three lakes are provided in Figure 3. Four fluorescent components were identified based on the DOM fluorescence characteristics of the lake sediments. Component C1 (Ex: 245 nm, Em: 430 nm) corresponded to a terrestrial humic-like substance [47]. Component C2 (Ex: 275/365 nm, Em: 455 nm) coincides with a fulvic acid-like substance known for its stability and resemblance to humus [48]. Components C3 (Ex: 280 nm, Em: 335 nm) and C4 (Ex: 320 nm, Em: 405 nm) conformed to protein tryptophan components and terrestrial humic acids, respectively [49,50]. Overall, C1, C2, and C4 are humic-like substances, whereas C3 is protein-like.

The fluorescence intensities and relative proportions of sediment DOM from various monitoring sites in the lake are shown in Figure 4. Erhai Lake exhibited the highest total average fluorescence intensity, followed by Dianchi Lake, with Yangzonghai Lake displaying the lowest average. It is worth noting that the northern region of Dianchi Lake exhibited a significantly higher total fluorescence intensity compared to the other areas of the lake. Moreover, the northern zone of Erhai Lake showed higher total fluorescence intensity than the southern and central zones. The total fluorescence intensity in Yangzonghai Lake decreased from the northern to the southern zones. The average proportions of the sum of C1, C2, and C4 in the sediments of Dianchi, Erhai, and Yangzonghai were 68.06%, 78.47%, and 66.08%, respectively, indicating that the sediment DOM in the three lakes mainly comprised humic-like substances. Humic-like components were more abundant in Caohai Lake compared to Waihai Lake, while in Waihai Lake, the concentration of humic-like substances displayed a declining gradient from south to north. In Erhai Lake, the northern region exhibited a significantly higher concentration of humic-like components compared to the central and southern regions. Additionally, a rise in humic-like material levels from north to south was confirmed in Yangzonghai Lake.

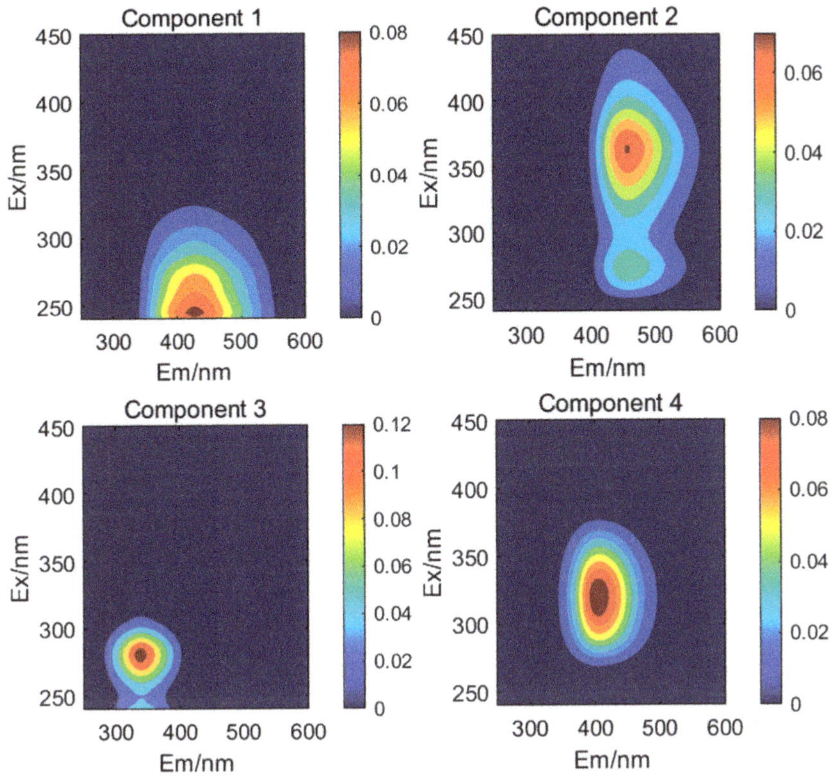

Figure 3. Fluorescence components of sediment DOM.

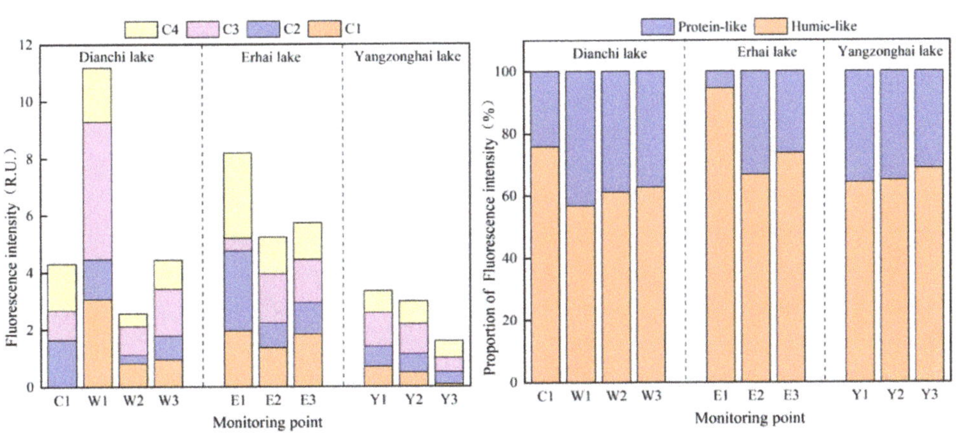

Figure 4. Fluorescence intensities and relative proportions of sediment DOM at each monitoring site.

3.4. Correlations Between Sediment P Fractions, DOM, and Human Activities

This study analyzed the Pearson correlation of DOM and human activities with sediment P fractions. As shown in Figure 5a, the DOM content was positively correlated with mobile-P ($p < 0.01$, $n = 10$), BD-P ($p < 0.01$, $n = 10$), and NaOH-nrP ($p < 0.05$, $n = 10$) in the sediments. A strong positive relationship was observed between the NH_4Cl-P and the terrestrial humic-like substance C1 component ($p < 0.001$, $n = 10$) and protein-like C3 component ($p < 0.01$, $n = 10$). As illustrated in Figure 5b, the indicators of human activity intensity were significantly and positively correlated with TP ($p < 0.001$, $n = 10$), immobile-P ($p < 0.001$, $n = 10$), NaOH-rP ($p < 0.01$, $n = 10$), and HCl-P ($p < 0.05$, $n = 10$) in the lake sediments.

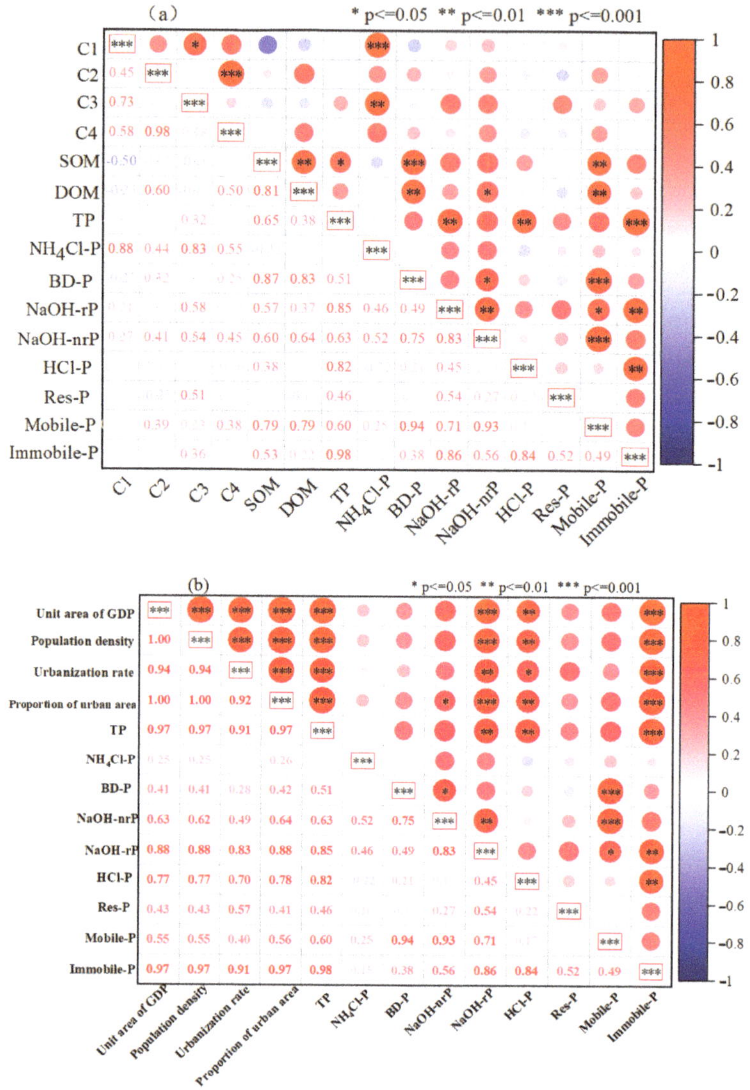

Figure 5. Correlation analysis of DOM and human activities with phosphorus. (**a**) The correlation coefficients between the P fractions and DOM characteristics. (**b**) The correlation coefficients between the P fractions and human activity indicators.

4. Discussion

4.1. The Influence of Sediment DOM on P Mobility

Despite its relatively low content in sediments, DOM plays a crucial role in influencing the potential mobility of phosphorus within sediments [51]. A significant positive correlation was found between the DOM and mobile-P contents in the sediments. Mobile-P, with its relatively short cycling timescale of approximately 10 years, is considered a critical contributor to lake eutrophication [13,52]. Among the various P-binding forms, DOM mainly influences the adsorption of BD-P in sediments and impacts the retention of NaOH-nrP.

In sediments, BD-P primarily consists of phosphorus bound to Fe oxides, with a smaller fraction associated with Mn oxides, and its release is predominantly controlled by redox conditions at the sediment–water interface [52]. In an anaerobic water environment, BD-P is released in high amounts from sediments and significantly affects the P concentration in the water. Previous studies indicated that BD-P is more likely to be retained in sediments in regions affected by eutrophication [53]. Therefore, BD-P often serves as a key indicator of endogenous P load in lake eutrophication scenarios [54,55]. This study showed that BD-P was the leading component of mobile-P in the sediments (53% on average), with a significant positive correlation observed between BD-P and the DOM content in the sediments. The significance of DOM in sediments is underscored, as it increases the potential for lake eutrophication by altering the interaction between Fe oxide and P in sediments. The BD-P content of the sediments was closely linked to the binding forms of Fe oxide. Typically, amorphous Fe oxide has a larger surface area and greater P adsorption capacity than crystalline Fe oxide [56]. Studies have shown that low-molecular-weight organic compounds produced via organic matter decomposition can form organic–metal chelates with stable chemical structures by bonding with metals, thereby augmenting P adsorption capacity [57].

NaOH-nrP primarily originates from the deposition of biological organisms [52,58,59]. A significant positive correlation was observed between the content of DOM and NaOH-nrP in the sediments. DOM is an important P reservoir in sediments; additionally, it may be related to the mechanisms of microbial action in the organophosphate cycle. Recent studies have found that the DOM humification process is conducive to the conversion of certain diester P species into RNA mononucleotide and β-glycerophosphoric acid, and inositol hexaphosphate can readily bind with polymer humus to form complex organophosphorus [25]. In addition, some studies have identified a significant negative correlation between the NaOH-nrP content and O/C and (O + N)/C [28]. In general, the higher ratios of O/C and (O + N)/C suggest that an increased number of functional groups, such as carbonyl, hydroxyl, carboxyl, amino, and nitro groups, are present in the aromatic core of DOM. Therefore, the humification process of DOM is likely to significantly affect the transformation and retention of organic P, and the specific underlying mechanisms warrant further investigation.

A strong positive correlation was found between the NH_4Cl-P content in the sediments and the terrestrial humic-like C1 component, as well as the protein-like C3 component. The adsorption of NH_4Cl-P in sediments is greatly influenced by the composition and structure of DOM. As NH_4Cl-P is weakly adsorbed in sediments that comprise P in interstitial water, P binds to $CaCO_3$ and separates from the degradation of bacterial cells in deposited plant debris [60,61]. The hydrophilic groups of the DOM compete with phosphate for adsorption or anion substitution, thereby inhibiting phosphate adsorption [62,63]. Previous studies have indicated that as the degree of humification of DOM increases, so does its molecular weight, whereas the number of functional groups such as hydroxyl, carboxyl, and carbonyl groups on aromatic nuclei decreases, creating a more conducive environment for the adsorption of NH_4Cl-P in sediments [64–66]. In addition, the protein-like component of DOM serves as the primary carrier of soluble organophosphorus. Microbial degradation of soluble organophosphorus directly influences P adsorption in interstitial water and sediment through a microbial regulation mechanism [67], subsequently modulating the NH_4Cl-P content in the sediments.

4.2. Human Activities Leading to P Retention in the Lake Sediments

Relevant studies have shown that human activities, including sewage discharge, land utilization, agricultural production, and phosphate mining, cause significant P pollution in sediments [68–70]. This study found that human activity intensity was significantly and positively correlated with TP, immobile-P, NaOH-rP, and HCl-P in the lake sediments.

This study revealed a direct correlation between the level of human activity in the lake basin and the TP concentration in the sediment, with higher intensity leading to increased sediment TP content. Higher population density and urbanization result in an elevated exogenous pollution load. As sinks for exogenous pollution, lake sediments inevitably accept substantial exogenous P loads, thereby augmenting the TP content of sediments. Among the three lakes in this study, Dianchi Lake had the highest sediment TP content, which was affected by the pollution intensity and pollution source characteristics (Section S3 in the Supplementary Material).

The level of human impact on lakes is positively correlated with the amount of immobile-P retained in lake sediments [71,72]. However, the intensity of human activities does not correlate with mobile-P because it displays strong mobility and bioavailability and is easily transformed [73]. Sediment immobile-P in the Dianchi, Erhai, and Yangzonghai lakes was mainly distributed in areas with marked human and agricultural activities. This phenomenon likely corresponded to the notable presence of NaOH-rP and HCl-P, which are easily influenced by human and agricultural activities [74,75].

A strong positive relationship existed between the proxy index measuring human activity intensity and NaOH-rP and HCl-P concentrations. NaOH-rP, an indicator of human activity, can be distinguished from industrial wastewater and domestic sewage [8]. Typically, P resides in urban lake sediments in the NaOH-rP form [76]. The NaOH-rP contents in Caohai Lake and Northern Waihai Lake were relatively high. After the 1960s, owing to the advancement of industrialization and urbanization, large portions of land in Caohai and the northern region of Waihai Lake in the Dianchi Lake Basin developed heavily, leading to an increase in population density [77]. Large volumes of industrial wastewater and domestic sewage are released into the lake, and NaOH-rP has been identified as a stable P form that tends to accumulate in sediments [78,79]. Compared with Dianchi Lake, the proportion of NaOH-rP in Erhai Lake and Yangzonghai Lake was relatively low, as they are categorized as non-urban lakes with a relatively lower population density, thus experiencing less impact from human activities [80,81]. The NaOH-rP in Erhai Lake was primarily distributed in the central zone. This is because the central zone is adjacent to Wase Town, where human activities and the introduction of Fe and Al have affected the region [82]. The spatial distribution of NaOH-rP in Yangzonghai Lake demonstrates a decreasing trend from the northern to the southern zone, likely because of the relatively high intensity of human activity in the northern zone of Yangzonghai. Domestic sewage is discharged into the lakes through the Baiyi River on the northern bank [83].

In contrast to NaOH-rP, HCl-P was widespread in the central and southern zones of Dianchi Lake. This was because the southern zone of Dianchi Lake is a geologically P-rich area with intensive mining and agricultural activities. Phosphorite extraction and fertilizer use have resulted in a significant influx of P into Dianchi Lake via runoff and erosion, eventually depositing into sediments [84]. HCl-P is the major P form in Erhai Lake, primarily distributed in the northern zone of Erhai Lake. The Miju River in the north contributes 51% of the water in Erhai Lake. Eryuan County, through which the Miju River flows, has a high population density and is mainly known for its agricultural practices, which result in significant agricultural nonpoint-source pollution. Agricultural runoff is the main source of P in Erhai Lake, with agricultural activities leading to increased HCl-P levels in sediments in the northern area of the lake [74]. Zhang et al. [56] showed that in lakes dominated by agriculture, the P forms in sediments are mainly HCl-P. The percentage of HCl-P in Yangzonghai Lake was lower than that of the Dianchi and Erhai lakes, indicating a decline from the northern region to the southern region. This is because the population and agriculture of Yangzonghai Lake are concentrated in the northern zone, facilitating the generation of HCl-P deposits.

4.3. The Impact of Lake Trophic and Ecological Types on Sediment P Mobility

Lakes in China are categorized based on the nutrients in the lake water, quantity of living biota, and primary producer species [85]. Currently, the coverage of aquatic vegetation in Dianchi Lake is <5%, and the algal biomass reaches the order of 10^7–10^8. This lake is a typical algal-type eutrophic shallow lake with an average depth of 4.4 m [44,61,86]. The coverage of aquatic vegetation in Erhai Lake is 13.2%, and the algal biomass reaches 10^7 orders of magnitude. Currently undergoing a delicate transition from grass-dominated to algae-dominated, the lake is a typical grass–algae mixed type with a medium nutrient level and an average depth of 10.5 m [22,61,87]. Yangzonghai Lake has moderate nutrient levels and a typical depth of 20.13 m [88]. Its aquatic vegetation coverage is low, and the algal biomass is approximately 10^6. The three lakes examined in this study exhibited high heterogeneity in terms of nutrient levels, aquatic biodiversity, richness, and water depth. This inevitably affects the retention and mobility of P in sediments.

Dianchi Lake exhibited the highest mobile-P content in its sediments, likely influenced by the nutrient levels within the lake. Organic materials are abundant in the sediments of eutrophic lakes. On the one hand, DOM is a P reservoir in sediments [89]. On the other hand, humus in DOM can form organic and inorganic complexes with Fe and Al, providing important adsorption sites for inorganic P and promoting P adsorption in sediments [90]. BD-P was the main component of mobile-P, indicating that high sediment DOM content was conducive to the adsorption of mobile-P. Liu et al. [41] studied Taihu Lake and showed that the biogeochemical cycles of P and OM are significantly influenced by Fe. The OM produced from macrophyte-dominated lakes has better aromatic properties, and Fe preferentially combines with highly aromatic compounds to form more OM-Fe-P and promotes the adsorption of P. Bao et al. [91] studied Chaohu Lake and noted that humic-like components could affect the dynamics of internal P through the formation of organic colloids and organic–inorganic ligands.

Erhai Lake and Yangzonghai Lake are mesotrophic lakes; however, Erhai Lake has a higher content and proportion of mobile-P in its sediments. Aquatic plants in lakes can affect the binding form of P in sediments by influencing its absorption and release [92]. Wang et al. [93] reported that synergistic interactions between aquatic plants and Fe-oxidizing bacteria can enhance BD-P formation and fixation in the sediment.

Water depth also has a significant influence on the form of P in the sediments. With increasing lake depth, there is a decrease in the presence of mobile-P in the sediment, indicating enhanced mobility of P in shallow lakes or the shallow parts of lakes. The richness and diversity of aquatic organisms are closely related to water depth, and shallow lakes or areas typically maintain greater biodiversity and richness. A recent study found that water depth influences phosphorus retention in sediments, primarily through its role in regulating ecosystem transitions [61]. Previous investigations have found a notable decline in P retention efficiency in sediments after a transition from a grass-dominated ecosystem to one dominated by algae [94]. In contrast, the potential mobile-P load of the sediment was significantly increased [52].

Specifically, the sampling point (C1) in Caohai Lake of Dianchi had the minimum water depth and maximum mobile-P content, which was 2–9 times higher than the mobile-P content of other sediment samples. The sampling point in the northern part of Waihai

Lake (W2) had a mobile-P content 1.7–4.4 times higher than that of other sediment samples. This disparity was attributed to cyanobacterial accumulation in the northern zone of Dianchi Lake, which was influenced by exogenous inputs from the primary urban area of Kunming in the northern and southeastern wind directions. Algal blooms lead to anaerobic conditions and the release of BD-P from the sediment. The breakdown of algae and the deposition of dead algal debris may result in the build-up of biogenic organic P, thereby improving the mobility of sediment P [95]. In contrast, the southern zone of Waihai Lake features a grassy lake with submerged vegetation historically, which may promote the transformation of mobile-P in the sediments into immobile-P [96]. Therefore, the mobile-P in the southern zone of Waihai Lake was relatively low.

Moreover, submerged vegetation was distributed in the northern and southern areas of Erhai Lake, whereas the central zone lacked vegetation because of water depth [97,98]. Mobile-P levels in Erhai Lake reached their highest at the center, whereas deeper lake depths hindered the mineralization and decomposition of NaOH-nrP. The central area of Erhai Lake had the smallest proportion of HCl-P compared with the other two sites because of the lack of submerged plants that helped accumulate HCl-P [99]. The central zone of Erhai Lake exhibited a notable depth with scarce submerged plant growth.

Yangzonghai Lake features Res-P as the primary form of P, which is mainly composed of macromolecular organophosphorus or other insoluble P [100]. Compared with eutrophic lakes, organic detrital P in oligotrophic lakes is easily transformed into insoluble P [101]. Compared to Dianchi Lake and Erhai Lake, the algae and aquatic plants in Yangzonghai Lake were less distributed, with deeper prevailing waters [102]. SOM primarily originates from adjacent rotten vegetation, which impedes its participation in P cycling. It directly combines with P to form residual P, which remains in the sediment. Therefore, the transition of lakes from being grass-dominated to being algae-dominated results in higher P mobility in the sediment [103].

4.4. Environmental Significance

This study investigated P-binding forms in the sediments of lakes with varying levels of human activity and trophic ecological types.

Lakes exhibiting high levels of human impacts and abundant algal density, such as Dianchi Lake, have a wide range of exogenous P sources, with algae serving as primary producers in relatively uncomplicated ecosystems [104]. The TP content in the sediments of Dianchi Lake was >2000 mg/kg. Moreover, Dianchi Lake revealed a reduced level of microbial abundance and diversity, thereby facilitating the retention of mobile-P with a content >200 mg/kg. However, elevated levels of mobile-P in sediments pose a high risk of release [43]. For this particular type of lake, it is advisable that, on the basis of external P load control, measures such as sediment dredging or in situ treatment should be taken to control the internal load to effectively manage the internal nutrient load by reducing P levels in the sediment and preventing its potential release from the overlying water.

A grass–algae mixed lake with moderate human activity intensity, such as Erhai Lake, has a low utilization rate of NH_4Cl-P, and NH_4Cl-P is easily left in the sediment, reaching an average content of 6.87 mg/kg. The middle part of Erhai Lake lacks aquatic plants because of the deep water depth, causing NH_4Cl-P in the sediment to be easily released into the water above, leading to a higher chance of eutrophication [105,106]. Submerged vegetation releases O through photosynthesis from the roots as it grows, changing the environment around the roots and causing Fe oxidation. Consequently, Fe oxide can attach to or absorb phosphate, decreasing the re-release of P [107]. Therefore, within the confines of Erhai Lake, which is characterized by a grass–algae ecosystem with uneven vegetation distribution, measures can be taken to restore aquatic vegetation and plant aquatic flora in areas with less vegetation distribution.

Lakes with moderate human activity intensity and mesotrophic algae types, such as Yangzonghai Lake, have less aquatic vegetation distribution, and Res-P is the main P-binding form in the sediments, accounting for 34% of the TP, posing challenges for its release into the overlying waters [108,109]. For lakes with a minimum endogenous pollution load and good water quality, emphasis should be placed on the prevention and management of exogenous pollution, restricting the inflow of nutrients such as N and P, and controlling algal reproduction.

There are significant variations in P morphology within lake sediments, which are influenced by differing intensities of human activity and distinct trophic and ecological types. Therefore, lake management strategies should be tailored to local conditions. To implement targeted scientific management measures, it is essential to conduct comprehensive investigations and measurements of water quality, sediments, and aquatic ecosystems. Moreover, it is crucial to consider the intensity of human activities and the pollution characteristics within watersheds.

5. Conclusions

This study investigated P-binding forms in the sediments of three plateau lakes in Yunnan Province—Dianchi Lake, Erhai Lake, and Yangzonghai Lake—which represent eutrophic algae-type, mesotrophic grass–algae mixed-type, and mesotrophic algae-type trophic states, respectively. The findings revealed significant impacts of sediment DOM, human activities, and lake ecosystem characteristics on P retention and mobility, providing new insights into the mechanisms driving internal phosphorus cycling in lakes.

Human activities were found to play a pivotal role in sediment P retention. Correlation analysis revealed strong positive correlations between lake sediment TP, immobile-P, NaOH-rP, and HCl-P with human activity. However, no significant correlation was found between human activity intensity and mobile-P, which is more dynamic and subject to transformation.

Lake ecosystem characteristics affect P mobility in sediments. The mobile-P content in the Dianchi Lake sediment was 40.6% and 278.3% higher than that in Erhai and Yangzonghai, respectively. The ecological shift from grass-dominated to algal-dominated systems increases P mobility because of the activity of microorganisms, algae, and aquatic plants. In addition, shallow water depths enhance P mobility, elevating the risk of internal pollution.

The relationship between sediment DOM and P dynamics is complex. This study showed a significant positive correlation between sediment DOM and mobile-P. On the one hand, DOM may facilitate the accumulation of BD-P through adsorption or hinder the formation of amorphous Fe oxide crystals. On the other hand, DOM acts as a carrier for organophosphorus, impacting microbial processes. This study revealed that the NH_4Cl-P content in the sediments was positively correlated with the C1 component of a terrestrial humic-like substance and with the C3 component of a protein-like substance. These findings highlight the critical role of DOM composition in shaping sediment P dynamics.

This study provides novel insights into how DOM, human activities, and lake trophic states affect sediment P retention and mobility, with implications for managing internal P loading in eutrophic lakes. The results emphasize the significant role of sediment DOM in modulating phosphorus dynamics, particularly mobile phosphorus forms, while human activities primarily promote the retention of phosphorus in sediments as immobile-P. This study highlights considerable variations in phosphorus forms across lake sediments, driven by human activity and trophic ecological types, underscoring the need for tailored management strategies to control internal phosphorus loading and mitigate eutrophication risks. Future research could explore the interplay between external nutrient inputs and internal P cycling to guide effective lake restoration strategies.

Supplementary Materials: The following supporting information can be downloaded at: https://www.mdpi.com/article/10.3390/toxics13020120/s1, Section S1: The basic characteristics of Dianchi Lake, Erhai Lake, and Yangzonghai Lake; Section S2: The SOM content and water quality of Dianchi Lake, Erhai Lake, and Yangzonghai Lake; Section S3: Effects of pollution intensity and pollution characteristics on TP content in the sediments of three lakes.

Author Contributions: X.W. conducted the experiments and wrote the manuscript. Y.W. conducted the experiments and analyzed the data. L.J. and J.H. modified the manuscript and provided funding support. H.Z. conducted the field investigation and made the figures in this paper. Z.H. conducted the field investigation and collected data. All authors have read and agreed to the published version of the manuscript.

Funding: This work was jointly supported by the National Major Science and Technology Program for Water Pollution Control and Treatment (No. 2012ZX07102-004) and Cultivating Plan Program for the Leader in Science and Technology of Yunnan Province (202305AD160060).

Institutional Review Board Statement: Not applicable.

Informed Consent Statement: Not applicable.

Data Availability Statement: The data will be made available upon request.

Conflicts of Interest: Author Zhengzheng Hao was employed by the company Yunnan Dianwei Environmental Protection Technology Co., Ltd. The remaining authors declare that the research was conducted in the absence of any commercial or financial relationships that could be construed as a potential conflict of interest.

References

1. Yan, Z.B.; Han, W.X.; Peñuelas, J.; Sardans, J.; Elser, J.J.; Du, E.Z.; Reich, P.B.; Fang, J.Y. Phosphorus accumulates faster than nitrogen globally in freshwater ecosystems under anthropogenic impacts. *Ecol. Lett.* **2016**, *19*, 1237–1246. [CrossRef]
2. Diaz, R.J.; Rosenberg, R. Spreading dead zones and consequences for marine ecosystems. *Science* **2008**, *321*, 926–929. [CrossRef]
3. Zhao, J.; Gao, Q.S.; Liu, Q.Q.; Fu, G. Lake eutrophication recovery trajectories: Some recent findings and challenges ahead. *Ecol. Indic.* **2020**, *110*, 105878. [CrossRef]
4. Ma, X.H.; Wan, H.B.; Zhou, J.; Luo, D.; Huang, T.; Yang, H.; Huang, C.C. Sediment record of polycyclic aromatic hydrocarbons in Dianchi lake, southwest China: Influence of energy structure changes and economic development. *Chemosphere* **2020**, *248*, 126015. [CrossRef]
5. Pan, F.; Guo, Z.R.; Cai, Y.; Fu, Y.Y.; Wu, J.Y.; Wang, B.; Liu, H.T.; Gao, A.G. Cyclical patterns and (im) mobilization mechanisms of phosphorus in sediments from a small creek estuary: Evidence from in situ monthly sampling and indoor experiments. *Water Res.* **2020**, *171*, 115479. [CrossRef] [PubMed]
6. Li, Q.M.; Zhang, W.; Wang, X.X.; Zhou, Y.Y.; Yang, H.; Ji, G.L. Phosphorus in interstitial water induced by redox potential in sediment of Dianchi Lake, China. *Pedosphere* **2007**, *17*, 739–746. [CrossRef]
7. Hu, M.P.; Liu, Y.M.; Zhang, Y.F.; Shen, H.; Yao, M.Y.; Dahlgren, R.A.; Chen, D.J. Long-term (1980–2015) changes in net anthropogenic phosphorus inputs and riverine phosphorus export in the Yangtze River basin. *Water Res.* **2020**, *177*, 115779. [CrossRef] [PubMed]
8. Yang, B.; Lan, R.Z.; Lu, D.L.; Dan, S.F.; Kang, Z.J.; Jiang, Q.C.; Lan, W.L.; Zhong, Q.P. Phosphorus biogeochemical cycling in intertidal surface sediments from the Maowei Sea in the northern Beibu Gulf. *Reg. Stud. Mar. Sci.* **2019**, *28*, 100624. [CrossRef]
9. Liang, Z.; Liu, Z.; Zhen, S.; He, R. Phosphorus speciation and effects of environmental factors on release of phosphorus from sediments obtained from Taihu Lake, Tien Lake, and East Lake. *Toxicol. Environ. Chem.* **2015**, *97*, 335–348. [CrossRef]
10. Long, Z.; Ji, Z.; Pei, Y. Characteristics and distribution of phosphorus in surface sediments of a shallow lake. *J. Environ. Sci.* **2023**, *124*, 50–60. [CrossRef] [PubMed]
11. Kong, M.; Ouyang, X.Y.; Han, T.L.; Wang, W.Z.; Yin, H.B.; Wang, Y. Combination of lanthanum-modified attapulgite and Vallisneria natans for immobilization of phosphorus in various types of sediments. *Chem. Eng. J.* **2024**, *492*, 152264. [CrossRef]
12. Liu, Q.; Liu, S.L.; Zhao, H.D.; Deng, L.; Wang, C.; Zhao, Q.H.; Dong, S.K. The phosphorus speciations in the sediments up-and down-stream of cascade dams along the middle Lancang River. *Chemosphere* **2015**, *120*, 653–659. [CrossRef] [PubMed]

13. Yin, H.; Zhang, M.; Yin, P.; Li, J. Characterization of internal phosphorus loading in the sediment of a large eutrophic lake (Lake Taihu, China). *Water Res.* **2022**, *225*, 119125. [CrossRef]
14. Duan, Z.; Tan, X.; Ali, I.; Wu, X.; Cao, J.; Xu, Y.; Shi, L.; Gao, W.; Ruan, Y.; Chen, C. Comparison of organic matter (OM) pools in water, suspended particulate matter, and sediments in eutrophic Lake Taihu, China: Implication for dissolved OM tracking, assessment, and management. *Sci. Total Environ.* **2022**, *845*, 157257. [CrossRef] [PubMed]
15. Toming, K.; Kotta, J.; Uuemaa, E.; Sobek, S.; Kutser, T.; Tranvik, L.J. Predicting lake dissolved organic carbon at a global scale. *Sci. Rep.* **2020**, *10*, 8471. [CrossRef]
16. Palansooriya, K.N.; Kim, S.; Igalavithana, A.D.; Hashimoto, Y.; Choi, Y.E.; Mukhopadhyay, R.; Sarkar, B.; Ok, Y.S. Fe (III) loaded chitosan-biochar composite fibers for the removal of phosphate from water. *J. Hazard. Mater.* **2021**, *415*, 125464. [CrossRef] [PubMed]
17. Wang, J.; Chen, Q.; Huang, S.; Wang, Z.; Li, D. Cyanobacterial organic matter (COM) positive feedback aggravates lake eutrophication by changing the phosphorus release characteristics of sediments. *Sci. Total Environ.* **2023**, *892*, 164540. [CrossRef] [PubMed]
18. Xing, X.; Chen, M.; Wu, Y.; Tang, Y.; Li, C. The decomposition of macrozoobenthos induces large releases of phosphorus from sediments. *Environ. Pollut.* **2021**, *283*, 117104. [CrossRef] [PubMed]
19. Liu, Y.; Zhu, Z.Q.; He, X.S.; Yang, C.; Du, Y.Q.; Huang, Y.D.; Su, P.; Wang, S.; Zheng, X.X.; Xue, Y.J. Mechanisms of rice straw biochar effects on phosphorus sorption characteristics of acid upland red soils. *Chemosphere* **2018**, *207*, 267–277. [CrossRef]
20. Kurek, M.; Harir, M.; Shukle, J.; Schroth, A.; Schmitt-Kopplin, P.; Druschel, G. Seasonal transformations of dissolved organic matter and organic phosphorus in a polymictic basin: Implications for redox-driven eutrophication. *Chem. Geol.* **2021**, *573*, 120212. [CrossRef]
21. Jiang, T.; Bravo, A.G.; Skyllberg, U.; Björn, E.; Wang, D.; Yan, H.; Green, N.W. Influence of dissolved organic matter (DOM) characteristics on dissolved mercury (Hg) species composition in sediment porewater of lakes from southwest China. *Water Res.* **2018**, *146*, 146–158. [CrossRef]
22. Ding, S.; He, J.; Liu, Y.; Jiao, L.X.; Zhao, H.C.; Cheng, Y.X. The adsorption-release behavior of sediment phosphorus in a typical "grass-algae" coexisting lake and its influence mechanism during the transition sensitive period. *Chemosphere* **2022**, *307*, 135903. [CrossRef] [PubMed]
23. Bai, L.L.; Cao, C.C.; Wang, C.L.; Wang, C.H.; Zhang, H.; Jiang, H.L. Roles of phytoplankton-and macrophyte-derived dissolved organic matter in sulfamethazine adsorption on goethite. *Environ. Pollut.* **2017**, *230*, 87–95. [CrossRef]
24. Lin, J.; Chen, X.; Liu, Y.L.; Wang, Y.B.; Shuai, J.X.; Chen, M.S. Fe/Mn (oxyhydr) oxides reductive dissolution promoted by cyanobacterial algal bloom-derived dissolved organic matter caused sediment W release during an algal bloom in Taihu Lake. *Water Res.* **2024**, *260*, 121899. [CrossRef] [PubMed]
25. Ni, Z.; Huang, D.; Wu, Y.; Li, Y.; Zhou, C.; Wang, S. Intrinsic linkage mechanisms of DOM properties to organic phosphorus in lake sediments: Evidence from coupled molecular weight ultrafiltration and spectral analysis. *Chem. Eng. J.* **2023**, *459*, 141496. [CrossRef]
26. Prüter, J.; Leipe, T.; Michalik, D.; Klysubun, W.; Leinweber, P. Phosphorus speciation in sediments from the Baltic Sea, evaluated by a multi-method approach. *J. Soils Sediments* **2020**, *20*, 1676–1691. [CrossRef]
27. Wen, S.; Liu, J.; Lu, Y.; Dai, J.; Huang, X.; An, S.; Jeppesen, E.; Liu, Z.; Du, Y. Composition regulates dissolved organic matter adsorption onto iron (oxy)hydroxides and its competition with phosphate: Implications for organic carbon and phosphorus immobilization in lakes. *J. Environ. Sci.* **2024**, *144*, 159–171. [CrossRef]
28. Zhao, Z.; Zhang, N.; Hu, X.; Bao, L.; Xia, Y. Effects of composition and structure of natural organic matter on phosphorus fractions in sediment from Lake Yangzonghai, Yunnan Province. *J. Lake Sci.* **2017**, *29*, 308–316.
29. Chen, Q.; Ni, Z.; Wang, S.; Guo, Y.; Liu, S.R. Climate change and human activities reduced the burial efficiency of nitrogen and phosphorus in sediment from Dianchi Lake, China. *J. Clean. Prod.* **2020**, *274*, 122839. [CrossRef]
30. Ni, Z.K.; Wang, S.R.; Cai, J.J.; Li, H.; Jenkins, A.; Maberly, S.C.; May, L. The potential role of sediment organic phosphorus in algal growth in a low nutrient lake. *Environ. Pollut.* **2019**, *255*, 113235. [CrossRef] [PubMed]
31. Cui, Y.; Wen, S.; Stegen, J.C.; Hu, A.; Wang, J. Chemodiversity of riverine dissolved organic matter: Effects of local environments and watershed characteristics. *Water Res.* **2023**, *250*, 121054. [CrossRef]
32. Fan, T.; Yao, X.; Sun, Z.; Sang, D.; Liu, L.; Deng, H.; Zhang, Y. Properties and metal binding behaviors of sediment dissolved organic matter (SDOM) in lakes with different trophic states along the Yangtze River Basin: A comparison and summary. *Water Res.* **2023**, *231*, 119605. [CrossRef]
33. Yang, H.; Yu, J.; Xu, W.; Wu, Y.; Lei, X.; Ye, J.; Geng, J.; Ding, Z. Long-time series ecological environment quality monitoring and cause analysis in the Dianchi Lake Basin, China. *Ecol. Indic.* **2023**, *148*, 110084. [CrossRef]
34. Luan, G.; Zhao, F.; Xia, J.; Huang, Z.; Feng, S.; Song, C.; Dong, P.; Zhou, X. Analysis of long-term spatio-temporal changes of plateau urban wetland reveals the response mechanisms of climate and human activities: A case study from Dianchi Lake Basin 1993–2020. *Sci. Total Environ.* **2024**, *912*, 169447. [CrossRef] [PubMed]

35. Zhang, Y.; Shi, T.R.; Yu, T.; Zhang, Y. Sediment Particle Size and the Distribution of Heavy Metals in the Typical Districts of Dianchi Lake. *Res. Environ. Sci.* **2013**, *26*, 370–379.
36. Zhang, Y.; Shen, J.; Feng, J.M.; Li, X.Y.; Liu, H.J.; Wang, X.Z. Composition, distribution, and source of organic carbon in surface sediments of Erhai Lake, China. *Sci. Total Environ.* **2023**, *858*, 159983. [CrossRef]
37. Zheng, W.; Zhang, E.; Langdon, P.G.; Wang, R. Systematic loss in biotic heterogeneity but not biodiversity across multiple trophic levels in Erhai lake, China. *Sci. Total Environ.* **2024**, *906*, 167479. [CrossRef]
38. Li, H.Y.; Zhang, H.; Chen, G.; Chang, F.; Duan, L.; Wang, J.; Lu, H.; Wu, H.; Hu, K. The Grain Size Distribution Characteristics of Surface Sediments from Plateau Lakes in Yunnan Province and Their Environmental Significances. *Acta Sedimentol. Sin.* **2017**, *35*, 499–507.
39. Yang, B.; Yang, S.; Wan, X.; Hu, H.; Hu, D.; Hua, M.; Liu, Y.; Pan, X. Temperature models for quantifying groundwater seepage flux applied in a deep lake of a plateau: Yangzonghai Lake, Yunnan, China. *Chemosphere* **2020**, *238*, 124674. [CrossRef]
40. Xiao, H.; Mao, C.; Wang, S.; Jia, Z.; Rao, W. Seasonal variation and provenance of organic matter in the surface sediments of the three gorges reservoir: Stable isotope analysis and implications for agricultural management. *Sci. Total Environ.* **2023**, *870*, 161886. [CrossRef] [PubMed]
41. Liu, J.J.; Zhang, Q.Y.; Chen, M.L.; Dai, J.R.; Gu, W.X.; Wen, S.L.; Du, Y.X. Composition of organic matter-iron-phosphorus associations in sediments of algae- and macrophyte-dominated zones in Lake Taihu. *Chem. Geol.* **2023**, *622*, 121375. [CrossRef]
42. Ma, J.; Yuan, Y.; Zhou, T.; Yuan, D.X. Determination of total phosphorus in natural waters with a simple neutral digestion method using sodium persulfate. *Limnol. Oceanogr. Methods* **2017**, *15*, 372–380. [CrossRef]
43. Saha, A.; Vijaykumar, M.; Das, B.; Samanta, S.; Khan, M.F.; Kayal, T.; Jana, C.; Chowdhury, A.R. Geochemical distribution and forms of phosphorus in the surface sediment of Netravathi-Gurupur estuary, southwestern coast of India. *Mar. Pollut. Bull.* **2023**, *187*, 114543. [CrossRef] [PubMed]
44. He, J.; Yang, Y.; Wu, X.; Zhi, G.Q.; Zhang, Y.; Sun, X.; Jiao, L.X.; Deng, W.M.; Zhou, H.B.; Shao, Z.; et al. Responses of dissolved organic matter (DOM) characteristics in eutrophic lake to water diversion from external watershed. *Environ. Pollut.* **2022**, *312*, 119992. [CrossRef] [PubMed]
45. Long, Y.; Hu, X.; Jiang, J.; Hu, J.; Zhu, C.; Zhou, S. Phosphorus sorption-desorption behaviors in the sediments cultured with *Hydrilla verticillata* and *Scripus triqueter* as revealed by phosphorus fraction and dissolved organic matter. *Chemosphere* **2021**, *271*, 129549. [CrossRef] [PubMed]
46. Yang, C.; Yang, P.; Geng, J.; Yin, H.; Chen, K. Sediment internal nutrient loading in the most polluted area of a shallow eutrophic lake (Lake Chaohu, China) and its contribution to lake eutrophication. *Environ. Pollut.* **2020**, *262*, 114292. [CrossRef] [PubMed]
47. Chen, M.; Kim, S.H.; Jung, H.J.; Hyun, J.H.; Choi, J.H.; Lee, H.J.; Huh, I.A.; Hur, J. Dynamics of dissolved organic matter in riverine sediments affected by weir impoundments: Production, benthic flux, and environmental implications. *Water Res.* **2017**, *121*, 150–161. [CrossRef] [PubMed]
48. Panettieri, M.; Guigue, J.; Prévost-Bouré, N.C.; Thévenot, M.; Lévêque, J.; Guillou, L.; Maron, P.A.; Santoni, A.L.; Ranjard, L.; Mounier, S. Grassland-cropland rotation cycles in crop-livestock farming systems regulate priming effect potential in soils through modulation of microbial communities, composition of soil organic matter and abiotic soil properties. *Agric. Ecosyst. Environ.* **2020**, *299*, 106973. [CrossRef]
49. Borisover, M.; Laor, Y.; Parparov, A.; Bukhanovsky, N.; Lado, M. Spatial and seasonal patterns of fluorescent organic matter in Lake Kinneret (Sea of Galilee) and its catchment basin. *Water Res.* **2009**, *43*, 3104–3116. [CrossRef]
50. Yamashita, Y.; Cory, R.M.; Nishioka, J.; Kuma, K.; Tanoue, E.; Jaffé, R. Fluorescence characteristics of dissolved organic matter in the deep waters of the Okhotsk Sea and the northwestern North Pacific Ocean. *Deep. Sea Res. Part II Top. Stud. Oceanogr.* **2010**, *57*, 1478–1485. [CrossRef]
51. Rapin, A.; Grybos, M.; Rabiet, M.; Mourier, B.; Deluchat, V. Phosphorus mobility in dam reservoir affected by redox oscillations: An experimental study. *J. Environ. Sci.* **2019**, *77*, 250–263. [CrossRef]
52. Ding, S.; Liu, Y.; Dan, S.F.; Jiao, L.X. Historical changes of sedimentary P-binding forms and their ecological driving mechanism in a typical "grass-algae" eutrophic lake. *Water Res.* **2021**, *204*, 117604. [CrossRef] [PubMed]
53. Wang, J.H.; Yang, C.; Dao, G.H.; Du, J.S.; Han, Y.P.; Wu, G.X.; Wu, Q.Y.; Hu, H.Y. Meteorological factors and water quality changes of Plateau Lake Dianchi in China (1990–2015) and their joint influences on cyanobacterial blooms. *Sci. Total Environ.* **2019**, *665*, 406–418. [CrossRef] [PubMed]
54. Ren, Z.; He, J.; Cheng, Q.; Ding, S.; Liu, W.; Duan, P.; Jiao, L.X. Climate change prior to human activity reduces the immobility of phosphorus in eutrophic alpine lake. *J. Clean. Prod.* **2022**, *335*, 130364. [CrossRef]
55. Shi, M.; Li, J.; Zhou, Q.; Wang, G.; Zhang, W.; Zhang, Z.; Gao, Y.; Yan, S. Interactions between elevated CO_2 levels and floating aquatic plants on the alteration of bacterial function in carbon assimilation and decomposition in eutrophic waters. *Water Res.* **2020**, *171*, 115398. [CrossRef] [PubMed]

56. Zhang, Y.; Zhang, T.; Huang, D.; Shao, D.; Luo, H. Geochemical and paleontological evidence of early Cambrian dynamic ocean oxygenation and its implications for organic matter accumulation in mudrocks at the Three Gorges area, South China. *Mar. Pet. Geol.* **2022**, *146*, 105958. [CrossRef]
57. Pontoni, L.; LaVecchia, C.; Boguta, P.; Sirakov, M.; D'Aniello, E.; Fabbricino, M.; Locascio, A. Natural organic matter controls metal speciation and toxicity for marine organisms: A review. *Environ. Chem. Lett.* **2022**, *20*, 797–812. [CrossRef]
58. Zhu, B.; Cheng, W. Impacts of drying–wetting cycles on rhizosphere respiration and soil organic matter decomposition. *Soil. Biol. Biochem.* **2013**, *63*, 89–96. [CrossRef]
59. Zhu, C.; Wang, Z.H.; Xue, B.; Yu, P.S.; Pan, J.M.; Wagner, T.; Pancost, R.D. Characterizing the depositional settings for sedimentary organic matter distributions in the Lower Yangtze River-East China Sea Shelf System. *Estuar. Coast. Shelf Sci.* **2011**, *93*, 182–191. [CrossRef]
60. Ding, X.; Guo, X.; Gao, H.; Gao, J.; Shi, J.; Yu, X.; Wu, Z. Seasonal variations of nutrient concentrations and their ratios in the central Bohai Sea. *Sci. Total Environ.* **2021**, *799*, 149416. [CrossRef]
61. Ren, H.; Fan, T.; Yao, X.; Ma, F.; Liu, L.; Ming, J.; Wang, S.; Zhang, Y.; Deng, H. Investigation of the variations in dissolved organic matter properties and complexations with two typical heavy metals under the influence of biodegradation: A survey of an entire lake. *Sci. Total Environ.* **2022**, *806*, 150485. [CrossRef]
62. Wang, H.B.; Liu, X.P.; Jin, B.J.; Shu, Y.C.; Sun, C.L.; Zhu, Y.G.; Lin, X.Y. High-molecular-weight dissolved organic matter enhanced phosphorus availability in paddy soils: Evidence from field and microcosm experiments. *Soil. Tillage Res.* **2024**, *240*, 106099. [CrossRef]
63. Zhang, T.; Chen, L.; Liu, X.; Shang, L.; Liu, Y.; Wang, C.; Zhao, S.; Chen, G. Spatial pattern and influencing factors of phytoplankton in lakes of central and southern Yunnan in summer. *Chin. J. Ecol.* **2020**, *39*, 3350–3362.
64. Wang, S.; Jin, X.; Zhao, H.; Zhou, X.; Wu, F. Effect of organic matter on the sorption of dissolved organic and inorganic phosphorus in lake sediments. *Colloids Surf. A* **2007**, *297*, 154–162. [CrossRef]
65. Wang, S.; Yi, W.; Yang, S.; Jin, X.; Wang, G.; Wu, F. Effects of light fraction organic matter removal on phosphate adsorption by lake sediments. *Appl. Geochem.* **2011**, *26*, 286–292. [CrossRef]
66. Wang, S.R.; Jin, X.C.; Zhao, H.C.; Zhou, X.N.; Wu, F.C. Effects of organic matter on phosphorus release kinetics in different trophic lake sediments and application of transition state theory. *J. Environ. Manag.* **2008**, *88*, 845–852. [CrossRef] [PubMed]
67. Du, S.T.; Shentu, J.L.; Luo, B.F.; Shamsi, I.H.; Lin, X.Y.; Zhang, Y.S.; Jin, C.W. Facilitation of phosphorus adsorption onto sediment by aquatic plant debris. *J. Hazard. Mater.* **2011**, *191*, 212–218. [CrossRef]
68. Jin, J.Q.; Xu, J.; Mo, Y.; Tang, H.; Wei, T.; Wang, Y.G.; Li, L. Response of sediments and phosphorus to catchment characteristics and human activities under different rainfall patterns with Bayesian Networks. *J. Hydrol.* **2020**, *584*, 124695. [CrossRef]
69. Martinez-Escobar, D.F.; Mallela, J. Assessing the impacts of phosphate mining on coral reef communities and reef development. *Sci. Total Environ.* **2019**, *692*, 1257–1266. [CrossRef] [PubMed]
70. Rashmi, K.; Karthika, T.; Roy, K.; Shinoji, A.; Kumawat, S.; Kala, R.P. *Agrochemicals in Soil and Environment: Impacts and Remediation*; Springer: Berlin/Heidelberg, Germany, 2022; pp. 313–345.
71. Tang, R.; Bai, J.; Zhang, L.; Wang, Y.; Liu, H.; Xia, J. Advances and hotspots of phosphorus form analysis of sediments based on different extraction methods using CiteSpace. *Ecohydrol. Hydrobiol.* **2024**, *24*, 884–892. [CrossRef]
72. Wen, S.; Zhong, J.; Li, X.; Liu, C.; Yin, H.; Li, D.; Ding, S.; Fan, C. Does external phosphorus loading diminish the effect of sediment dredging on internal phosphorus loading? An in-situ simulation study. *J. Hazard. Mater.* **2020**, *394*, 122548. [CrossRef] [PubMed]
73. Ye, H.; Yuan, X.; Han, L.; Yin, H.; Jin, J. Comparison of phosphorus fraction distribution and influencing factors of suspended and surface sediments in the Tiaoxi watershed, China. *Water Sci. Technol.* **2017**, *75*, 2108–2118. [CrossRef] [PubMed]
74. Dittrich, M.; Chesnyuk, A.; Gudimov, A.; McCulloc, J.; Quazi, S.; Young, J.; Winter, J.; Stainsby, E.; Arhonditsis, G. Phosphorus retention in a mesotrophic lake under transient loading conditions: Insights from a sediment phosphorus binding form study. *Water Res.* **2013**, *47*, 1433–1447. [CrossRef] [PubMed]
75. Wang, S.; Zhang, M.; Li, B.; Xing, D.; Wang, X.; Wei, C.; Jia, Y. Comparison of mercury speciation and distribution in the water column and sediments between the algal type zone and the macrophytic type zone in a hypereutrophic lake (Dianchi Lake) in Southwestern China. *Sci. Total Environ.* **2012**, *417*, 204–213. [CrossRef] [PubMed]
76. Zhang, Z.; Jiang, L.; Chen, M.; Li, J.; Zhang, L.; Zhang, J.; Liao, N. Release and transformation of phosphorus in sediment following seasonal freezing-thawing cycles. *J. Contam. Hydrol.* **2022**, *247*, 103978. [CrossRef] [PubMed]
77. Wang, J.; Mu, X.; Chen, S.; Liu, W.; Wang, Z.; Dong, Z. Dynamic evaluation of water resources carrying capacity of the Dianchi Lake Basin in 2005–2015, based on DSPERM framework model and simulated annealing-projection pursuit model. *Reg. Sustain.* **2021**, *2*, 189–201. [CrossRef]
78. Wang, S.; Wang, J.; Li, M.; Du, F.; Yang, Y.; Lassoie, J.P.; Hassan, M.Z. Six decades of changes in vascular hydrophyte and fish species in three plateau lakes in Yunnan, China. *Biodivers. Conserv.* **2013**, *22*, 3197–3221. [CrossRef]

79. Yan, K.; Yuan, Z.; Goldberg, S.; Gao, W.; Ostermann, A.; Xu, J.; Zhang, F.; Elser, J. Phosphorus mitigation remains critical in water protection: A review and meta-analysis from one of China's most eutrophicated lakes. *Sci. Total Environ.* **2019**, *689*, 1336–1347. [CrossRef] [PubMed]
80. Han, X.; Feng, L.; Chen, X.; Yesou, H. MERIS observations of chlorophyll-a dynamics in Erhai Lake between 2003 and 2009. *Int. J. Remote Sens.* **2014**, *35*, 8309–8322. [CrossRef]
81. Xie, G.; Nu, W.; Yang, S. Land use and water quality changes in Yangzonghai Lake, Songhuaba Reservoir and Yunlong Reservoir basins. *Water Resour. Prot.* **2007**, *5*, 15–17.
82. Rauch, J.N. Global spatial indexing of the human impact on Al, Cu, Fe, and Zn mobilization. *Environ. Sci. Technol.* **2010**, *44*, 5728–5734. [CrossRef]
83. Lin, J.; Sun, Q.; Ding, S.; Wang, D.; Wang, Y.; Chen, M.; Shi, L.; Fan, X.; Tsang, D.C. Mobile phosphorus stratification in sediments by aluminum immobilization. *Chemosphere* **2017**, *186*, 644–651. [CrossRef] [PubMed]
84. Zhou, Z.; Wang, Y.; Yang, H.; Liu, A.; Wu, S.; Teng, H.; Niu, X.Y. Sedimentary record of nutrients and sources of organic matter in the Shuanglong reservoir, Dianchi watershed, China. *Environ. Sci. Pollut. Res.* **2021**, *28*, 17763–17774. [CrossRef]
85. Klamt, A.M.; Qian, F.; Hu, K.; Wang, J.; Huang, L.; Li, R.; Chen, G. The rise and fall of primary producers and consumers in a multiply-stressed shallow lake (Lake Qilu, China) over the last 200 years. *Ecol. Indic.* **2021**, *129*, 107891. [CrossRef]
86. He, L.; Chen, G.; Wang, X.; Shen, J.; Zhang, H.; Lin, Y.; Shen, Y.; Lang, F.; Gong, C.L. Pollution Characteristics and Risk Assessment of Heavy Metals in the Sediments of the Inflow Rivers of Dianchi Lake, China. *Toxics* **2024**, *12*, 322. [CrossRef]
87. Zhu, R.; Wang, H.; Chen, J.; Shen, H.; Deng, X.W. Use the predictive models to explore the key factors affecting phytoplankton succession in Lake Erhai, China. *Environ. Sci. Pollut. Res.* **2018**, *25*, 1283–1293. [CrossRef]
88. Pan, Y.H.; Wang, H.B.; Gu, Z.P.; Xiong, G.H.; Yi, F. Accumulation and translocation of heavy metals by macrophytes. *Acta Ecol. Sin.* **2010**, *30*, 6430–6441.
89. Zhu, Y.; Wu, F.; He, Z.; Giesy, J.P.; Feng, W.; Mu, Y.; Feng, C.; Zhao, X.; Liao, H.; Tang, Z. Influence of natural organic matter on the bioavailability and preservation of organic phosphorus in lake sediments. *Chem. Geol.* **2015**, *397*, 51–60. [CrossRef]
90. Takahashi, T.; Dahlgren, R.A. Nature, properties and function of aluminum–humus complexes in volcanic soils. *Geoderma* **2016**, *263*, 110–121. [CrossRef]
91. Bao, Y.; Huang, T.; Ning, C.W.; Sun, T.T.; Tao, P.L.; Wang, J.; Sun, Q.Y. Changes of DOM and its correlation with internal nutrient release during cyanobacterial growth and decline in Lake Chaohu, China. *J. Environ. Sci.* **2023**, *124*, 769–781. [CrossRef] [PubMed]
92. Wu, W.; Yuan, L.; Li, E.; Liu, W.; Li, W. Effects of Aquatic Macrophyte on Phosphorus in Lake Sediment. *Hubei Agric. Sci.* **2007**, *46*, 1031–1034.
93. Wang, J.; Gao, M.; Yang, Y.; Lu, S.; Wang, G.; Qian, X.Q. Interactions of vallisneria natans and iron-oxidizing bacteria enhance iron-bound phosphorus formation in eutrophic Lake sediments. *Microorganisms* **2022**, *10*, 413. [CrossRef]
94. Cheng, Y.X.; Jiao, L.X.; Cheng, Q.L.; He, J.; Zhang, Y.; Ding, S. The evolution of a typical plateau lake from macrophyte to algae leads to the imbalance of nutrient retention. *Water Res.* **2023**, *236*, 119937. [CrossRef] [PubMed]
95. Yin, H.; Yin, P.; Yang, Z. Seasonal sediment phosphorus release across sediment-water interface and its potential role in supporting algal blooms in a large shallow eutrophic Lake (Lake Taihu, China). *Sci. Total Environ.* **2023**, *896*, 165252. [CrossRef]
96. Zhang, Y.; Zhang, C.; Qin, J.; Chen, Z.; Chen, Y.; Li, J.; Wang, X. Introducing La $(OH)_3$ nanoparticles into attapulgite for the control of sediments internal phosphorus release: Effectiveness, mechanisms, and microbial community response. *J. Environ. Chem. Eng.* **2023**, *11*, 110260. [CrossRef]
97. Wang, S.R.; Zhang, L.; Ni, L.Y.; Zhao, H.C.; Jiao, L.X.; Yang, S.W.; Guo, L.G.; Shen, J.Z. Ecological degeneration of the Erhai Lake and prevention measures. *Environ. Earth Sci.* **2015**, *74*, 3839–3847. [CrossRef]
98. Yang, S.; Yu, X.L.; Gao, D.; Zhang, K.; Wu, J.; Zhang, X.; Cui, Y. Health Assessment on Lake Erhai Using Benthic Index of Biotic Integrity. *Environ. Sci. Technol.* **2023**, *46*, 181–188.
99. Liu, C.; Du, Y.; Zhong, J.; Zhang, L.; Huang, W.; Han, C.; Chen, K.; Gu, X. From macrophyte to algae: Differentiated dominant processes for internal phosphorus release induced by suspended particulate matter deposition. *Water Res.* **2022**, *224*, 119067. [CrossRef]
100. Liu, S.; Wang, J.; Lin, C.; He, M.; Liu, X. Geochemical baseline level and function and contamination of phosphorus in Liao River Watershed sediments of China. *J. Environ. Manag.* **2013**, *128*, 138–143. [CrossRef]
101. Wilson, T.A.; Amirbahman, A.; Norton, S.A.; Voytek, M.A. A record of phosphorus dynamics in oligotrophic lake sediment. *J. Paleolimnol.* **2010**, *44*, 279–294. [CrossRef]
102. Zheng, X.; Chen, L.; Qiu, F.; Zhang, T.; Zhang, Z.; Shang, L.; Bai, N.; Chen, X.; Zhao, S.; Kong, L.; et al. Spatio-temporal distribution and driving factors of phytoplankton biomass in Lake Yangzong under the background of arsenic pollution treatment. *J. Lake Sci.* **2024**, *36*, 34–42.
103. Zhang, J.; Zuo, A.; Salimova, L.; Aijun, L.; Ling, L. Phytoplankton distribution characteristics and its relationship with bacterioplankton in Dianchi Lake. *Environ. Sci. Pollut. Res.* **2020**, *27*, 40592–40603. [CrossRef]

104. Li, G.; Li, L.; Pan, M.; Xie, Z.; Li, Z.; Xiao, B.; Liu, G.; Chen, J.; Song, L. The degradation cause and pattern characteristics of Lake Dianchi ecosystem and new restoration strategy of ecoregion and step-by-step implementation. *J. Lake Sci.* **2014**, *26*, 485–496.
105. Chu, Z.; Ye, B.; Tian, G.; Pang, Y.; Hu, X. Spatial distribution characteristics and estimation of submerged plant biomass in Lake Erhai. *Res. Environ. Sci.* **2014**, *27*, 1–5.
106. Tian, Y.; Chen, X.; Lv, C.; Shan, H.; Chou, Q.; Lv, X.; Zhang, X.; Ni, L.; Cao, T. Diversity and distribution status of aquatic plants in the lakeshore zone of Lake Erhai. *J. Lake Sci.* **2023**, *35*, 941–949.
107. Xu, D.; Yan, P.; Liu, Z.; Zhang, M.; Yan, W.; Liu, Y.; Wu, Z.; Zhang, Y. Spatial distribution of phosphorus forms and the release risk of sediments phosphorus in West Lake, Hangzhou, China. *Ecol. Eng.* **2021**, *173*, 106421. [CrossRef]
108. Tao, J.; Chen, G.; Chen, X.; Chen, L.; Huang, L.; Liu, Y.; Shi, H.; Hu, K.; Wang, J.; Kang, W. Long-term pattern of diatom community responses to water pollution and hydrological regulation in Yangzong Lake. *Geogr. Res.* **2016**, *35*, 1899–1911.
109. Zhang, H.; Li, Q.; Zhang, X.; Chen, W.; Ni, J.; Yang, L.; Wei, R. Insight into the mechanism of low molecular weight organic acids-mediated release of phosphorus and potassium from biochars. *Sci. Total Environ.* **2020**, *742*, 140416. [CrossRef] [PubMed]

toxics

Article

Per- and Poly-Fluoroalkyl Substances, and Organophosphate Flame Retardants in the Upper Yangtze River: Occurrence, Spatiotemporal Distribution, and Risk Assessment

Wen Sun [1,2], Zhiyou Fu [2], Yueyue Liu [2], Yingchen Bai [2], Yuyan Zhao [1], Chen Wang [2,*] and Fengchang Wu [2,*]

[1] College of Geo-Exploration Science and Technology, Jilin University, Changchun 130026, China; sunwen0039@163.com (W.S.); zhaoyuyan@jlu.edu.cn (Y.Z.)

[2] State Key Laboratory of Environmental Criteria and Risk Assessment, Chinese Research Academy of Environmental Sciences, Beijing 100012, China; fuzy@craes.org.cn (Z.F.); liuyueyue0103@mail.bnu.edu.cn (Y.L.); baiyc@craes.org.cn (Y.B.)

* Correspondence: wangchen@craes.org.cn (C.W.); wufengchang@vip.skleg.cn (F.W.)

Abstract: Contaminants of Emerging Concern (CECs), including per- and polyfluoroalkyl substances (PFASs) and organophosphate flame retardants (OPFRs), have raised global concerns due to their persistence, bioaccumulation potential, and toxicity. This study presents a comprehensive investigation of the occurrence, spatiotemporal distribution, potential sources, and the ecological and human health risks associated with 18 PFASs and 9 OPFRs in the surface waters of the upper Yangtze River, China. The water samples were collected from the main stream and five major tributaries (Min, Jinsha, Tuo, Jialing, and Wu Rivers) in 2022 and 2023. The total concentration of PFASs and OPFRs ranged from 16.07 to 927.19 ng/L, and 17.36 to 190.42 ng/L, respectively, with a consistently higher concentration observed in the main stream compared to the tributaries. Ultra-short-chain PFASs (e.g., TFMS) and halogenated OPFRs (e.g., TCPP) were the predominant compounds, likely originating from industrial discharges, wastewater effluents, and other anthropogenic sources. Ecological risk assessments indicated low-to-moderate risks at most sampling sites, with higher risks near wastewater discharge points. Human health risk evaluations suggested negligible non-carcinogenic risks but identified potential carcinogenic risks from OPFR exposure for adults at specific locations, particularly in Leshan city. This study highlights the importance of understanding the fate and impacts of PFASs and OPFRs in the upper Yangtze River, and provides valuable insights for developing targeted pollution control strategies and risk management measures.

Keywords: Contaminants of Emerging Concern (CECs); surface water; spatiotemporal distribution; risk assessment

1. Introduction

Contaminants of Emerging Concern (CECs), encompassing a wide array of anthropogenic compounds such as pharmaceuticals, personal care products, and endocrine-disrupting chemicals, have garnered significant attention in recent years due to their ubiquitous presence in the environment and potential adverse effects on ecosystems and human health [1,2]. Among these, per- and polyfluoroalkyl substances (PFASs) and organophosphate flame retardants (OPFRs) have emerged as two important classes, drawing particular attention due to their unique properties and widespread use [3,4]. PFASs are used in a variety of consumer products (e.g., food packaging, surfactants, pesticides, and non-stick coat-

 https://doi.org/10.3390/toxics13020116

ings) due to their unique chemical properties (e.g., surface activity, acid resistance, chemical stability, heat resistance, water and oil repellency) [5–7]. Additionally, PFASs are often found in the environment because their chemical structure makes them highly resistant to degradation through processes like hydrolysis, photolysis, and biodegradation [8–10]. As an alternative to bromine-based flame retardants, OPFRs have gained widespread adoption as flame retardants and plasticizers in various industrial products [11,12]. However, unlike their brominated counterparts, OPFRs do not chemically polymerize with the substrates to which they are applied, which facilitates the release of OPFRs into the environment during the production, use, and disposal of treated products [12,13]. The widespread use of PFASs and OPFRs, coupled with their continuous emissions and the inadequacy of current wastewater treatment technologies, has resulted in the ubiquitous presence of these contaminants in surface waters worldwide [14–19]. This highlights the need for a comprehensive understanding of the occurrence, fate, transport, and effects of these contaminants in aquatic environments, as well as the development of effective strategies for their mitigation and remediation to protect ecosystem and human health [20,21].

The ubiquitous presence of PFASs and OPFRs in aquatic environments, coupled with growing evidence of their adverse effects on organisms, has raised growing significant concerns. Studies have linked PFAS exposure to a range of health issues, including immune dysfunction, metabolic disorders, reproductive toxicity, and an increased risk of various cancers [22]. In response to these findings, international agencies such as the United States Environmental Protection Agency (US EPA) and the European Commission have implemented restrictions on the use of PFASs to mitigate their environmental and human health impacts [23]. Similarly, OPFRs have raised significant concerns due to their structural similarities with organophosphate pesticides, which are known for their multiple toxicities, including neurotoxicity [24]. Research has associated OPFRs with various negative health outcomes, such as genotoxicity, endocrine disruption, immunotoxicity, neurotoxicity, and potential carcinogenicity [25,26]. Reflecting these concerns, the European Union has taken proactive measures by banning several OPFRs in children's products due to their potential carcinogenic risks [27]. Given the growing evidence highlighting the adverse effects of PFASs and OPFRs, the long-term monitoring of these CECs in aquatic environments is crucial for understanding their environmental fate, transport, and impacts, as well as for informing the development of effective regulatory measures and remediation strategies to safeguard ecosystem and human health.

The Yangtze River Basin, the largest river basin in Eurasia, plays a vital role in supporting the livelihoods of hundreds of millions of people in China and is one of the country's most economically developed regions [28]. However, in recent decades, the basin, particularly its upper reaches, has experienced significant pollution [29]. While several studies have investigated the occurrence and concentration of PFASs and OPFRs in Yangtze River [30–33], most have focused on the middle and lower reaches, leaving a critical knowledge gap regarding the presence and impacts of these contaminants in the surface water of the upper reaches. Chongqing and Sichuan, two key provincial regions in the upper Yangtze River, host rapidly developing industries and are major consumers of chemicals and pharmaceuticals. These activities likely contribute to heightened pollution from PFASs and OPFRs in the basin, further emphasizing the need for a comprehensive understanding of their environmental fate and impacts. Given the growing concerns about the adverse effects of PFASs and OPFRs on ecosystem and human health, and the limited information available on their occurrence in the upper Yangtze River, it is crucial to evaluate the pollution status and conduct a comprehensive risk assessment of these contaminants in this region.

To address the knowledge gaps and growing concerns surrounding PFASs and OPFRs in the upper Yangtze River, this study systematically investigated the occurrence, spatiotemporal distribution, and potential sources of 18 PFASs and 9 OPFRs in water bodies of the region, including the main stream and four tributaries. By employing advanced analytical techniques and comprehensive sampling strategies, this research aimed to provide a deep understanding of the environmental behavior and fate of these contaminants in this ecologically and economically important region. Furthermore, a comprehensive risk assessment was conducted to evaluate the potential ecological and human health risks associated with PFASs and OPFRs in the upper Yangtze River. The findings of this study contribute to the development of a robust risk assessment framework for these contaminants in China's water environment, serving as a scientific foundation for informed decision-making and effective pollution control strategies to safeguard the environment and human health in Yangtze River.

2. Materials and Methods

2.1. Chemicals and Reagents

A total of 18 PFASs and 9 OPFRs were monitored in this study. The details of the target compounds, including their acronyms, CAS numbers, empirical formulas, dissociation constants(pKa), Log K_{ow} values, and common uses, are provided in Supplementary Materials (Text S1, Table S1). Mass-labeled compounds containing ^{2}H and ^{13}C isotopes were used as the internal standards (IS). All standards were purchased from J&K Scientific Ltd. (Beijing, China) and Sigma-Aldrich (St. Louis, MO, USA). Ultrapure water was obtained from a Milli-Q water purification system (Millipore, Bedford, MA, USA). HPLC-MS-grade methanol, acetonitrile, ammonia, and dichloromethane were purchased from Fisher Chemical, Inc. (Waltham, MA, USA). LC-MS-grade ammonium acetate and formic acid ($\geq$95%) were purchased from Sigma-Aldrich Inc. (St. Louis, MO, USA). Solid-phase extraction (SPE) cartridges Oasis HLB (500 mg, 6 cc) and Oasis WAX (500 mg, 6 cc) were purchased from Waters Inc. (Milford, MA, USA).

2.2. Study Area and Sample Collection

The Sichuan and Chongqing reaches of Yangtze River were selected as the study area. Water samples were collected from both the main channel of Yangtze River and its key tributaries, including the Min River, Jinsha River, Tuo River, Jialing River, and Wu River. Water samples were collected from the upper Yangtze River during two sampling campaigns in August 2022 (dry season) and August 2023 (flood season). In August 2022, a total of 19 samples were collected, including 13 samples from stream sites and 6 samples from wastewater treatment plant (WWTP) effluent sites. The sampling campaign was expanded in August 2023, with a total of 23 samples collected, comprising 17 stream samples and 6 WWTP effluent samples (Figure 1). For each sampling point, three individual samples (n = 3) were combined and thoroughly mixed to prepare a representative sample. Detailed information on the sampling locations is provided in Table S2. The samples were collected in amber glass bottles (4 L) that were pre-cleaned and rinsed with tap water, methanol, Milli-Q water, and water at the sampling site. The samples were stored and transported at 4 °C protected from light, and extracted within 24 h.

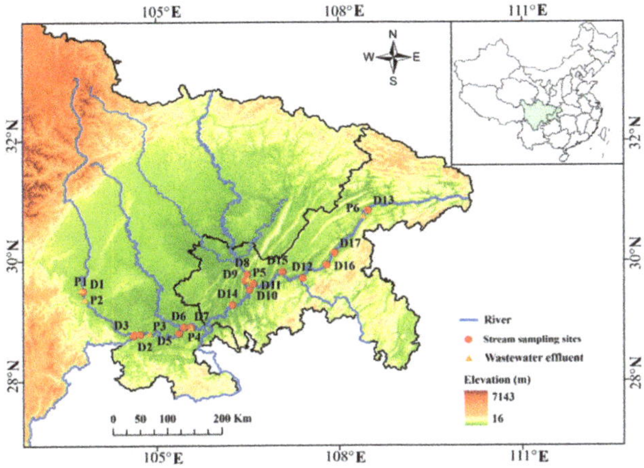

Figure 1. Sampling locations in the Sichuan and Chongqing Reach of Yangtze River. Red dots represent stream sampling sites, and yellow triangles represent wastewater effluent. There were 13 surface water sampling sites (D1–D13) along 2022, and 17 surface water sampling sites (D1–D17) along the 2023.

DZB-718L portable multiparameter water quality meter (Leici Inc., Shanghai, China) was used to measure water temperature (T), pH, oxidation–reduction potential (ORP), conductivity (Cond), dissolved oxygen (DO) at each site. The water quality parameters are summarized in Table S2.

2.3. Sample Preparation, Extraction, and Analysis

Samples (500 mL each) were filtered through 0.45 μm glass microfiber filters (Whatman GF/A) and stored at −20 °C until further processing and analysis. Targeted analyses were conducted using the methods described in previous studies [34–36]. Briefly, the filtered water samples were spiked with 50 ng of mixed internal standards and then enriched using activated SPE cartridges, and eluted with an appropriate solvent. The eluates were then concentrated under a stream of nitrogen (N_2), and reconstituted in methanol to a final volume of 500 μL (detailed in Text S2).

The target analytes were quantified using an ultra-high-performance liquid chromatography system (HPLC, 1290, Agilent, Santa Clara, CA, USA) coupled to a triple quadrupole mass spectrometer (MS/MS, 6460A, Agilent, Santa Clara, CA, USA) equipped with an electrospray ionization (ESI) interface (Agilent Technologies, Santa Clara, CA, USA). Data analysis, both qualitative and quantitative, was performed using MassHunter workstation software (GL Sciences, Tokyo, Japan). The injection volume for each sample was 5 μL. Detailed instrumental parameters and analytical methods are provided in Table S3.

2.4. Quality Assurance and Quality Control

Isotopically labeled internal standards and matrix-matched calibration were used to quantify target compounds [37]. The calibration curves demonstrated excellent linearity, with correlation coefficients (R^2) $\geq$ 0.99 for the concentration ranges of 1–200 ng/mL for PFASs and 1–500 ng/mL for OPFRs. The limits of detection (LOD) and quantification (LOQ) were determined based on the signal/noise ratios of 3 and 10, respectively. Each group (3–5 samples) contained a blank sample of which the concentration was always below the LOD. Detailed information on LOD, LOQ, and the recoveries of target compounds is shown in Table S4.

2.5. Annual Flux Estimation

The annual emission flux of target pollutants in the upper reaches of Yangtze River is estimated by the following formula [38]:

$$F = C \times Q \times 10^{-12} \tag{1}$$

where the flux is expressed as F (tons/year), C is the concentration of pollutants in the water phase (ng/L), and Q is the annual runoff of the basin. The contribution of tributaries to the main stream of Yangtze River can be estimated. The annual average flow data of the upper Yangtze River main stream and its tributaries are from the Statistical Yearbook (Table S5).

2.6. Ecological Risk Assessment

The ecological risk assessment was based on the European Commission Guidelines (EC, 2003) and recent research methods [39], and risk quotient (RQ) value was estimated by the following formula:

$$RQ = \frac{MEC}{PNEC} \tag{2}$$

where, MEC refers to measured environmental concentration and PNEC refers to the predicted no-effect concentration of the analyte.

The ecological risk of the mixtures was assessed using the following formula [40]:

$$RQ_{STU} = \sum\nolimits_{i}^{n} \frac{MEC_i}{\min\left(PNEC_{i,algae}, PNEC_{i,crustacean}, PNEC_{i,fish}\right)} \tag{3}$$

where STU refers to the sum of Toxic Units, and MEC_i is the measured environmental concentration of the *i*th component in an n-compound mixture.

Details of PNECs and RQs are provided in Table S6. Preliminary risk assessment level of pollutants: no risk ($\leq$0.01), low risk (0.01–0.1), medium risk (0.1–1), or high risk (>1) [41].

2.7. Human Health Risk Assessment

Health risk assessments for human exposure to chemicals typically include both non-carcinogenicity (non-CR) and carcinogenicity (CR) risk [31]. The hazard index (HI) is usually used to assess the non-CR of a compound. The average daily intake (ADI) of PFASs and OPFRs through drinking water was estimated and compared with the reference dose (RfD) and slope factors (SFs) to derive the HI and CR values. The total HI and CR were estimated by summing the studied compounds species HI and CR [31,42]. ADI, HI, and CR were calculated using the following formula:

$$ADI_i = \frac{C_i \times IR_i \times EF \times ED}{AT \times BW} \tag{4}$$

$$HI_i = \frac{ADI_i}{RfDi} \tag{5}$$

$$HI = \sum\nolimits_{i}^{n} HI_i \tag{6}$$

$$CR_i = ADI_i \times SF_i \tag{7}$$

$$CR = \sum\nolimits_{i}^{n} CR_i \tag{8}$$

where C_i is the concentration of each PFAS and OPFR in water (ng/L), IR_i is the intake rate of water (L/day), EF is the exposure frequency (day/year), ED is the exposure duration (year), AT is the average time (days), and BW is the average body weight (kg). RfD_i is the reference dose for each PFASs and OPFRs (ng/kg/day), and SF_i is the slope factors for each PFASs and OPFRs via ingestion.

The parameter values used in the health risk assessment are provided in Tables S7 and S8. Generally, HI < 1 is considered to be free of adverse health effects, whereas HI > 1 indicates the possibility of non-CR exceeding acceptable levels [43]. CR < 10^{-6} was considered negligible, and 10^{-6} < CR < 10^{-4} and CR > 10^{-4} indicated potential and high carcinogenic risks, respectively [44].

2.8. Statistical Analysis

SPSS 19.0 and Origin 2021 were used for data analysis. ArcGIS 10.7 was used for drawing the map of the sample sites and analyzing the spatial variation in PFAS and OPFR concentration in the upper Yangtze River. Nonmetric multidimensional scaling (NMDS) and heatmaps were produced with R statistical software (version 3.5.1). The Spearman correlation analysis was used to examine the possible correlations among various PFASs and OPFRs in the samples.

3. Results and Discussion

3.1. Occurrence and Concentration of PFASs and OPFRs in the Upper Yangtze River

The concentration of PFASs and OPFRs in surface water and effluent wastewater (minimum, maximum, median, mean, and detection frequency) are summarized in Table 1 and shown in Figure 2A,B. In surface water, the total PFAS concentration ranged from 16.07 to 454.68 ng/L in 2022, and from 23.13 to 1231.11 ng/L in 2023. The concentrations in 2023 were higher than those in 2022, both for individual compounds and for total PAFSs (Figure 2A). Among the 18 PFASs analyzed, 6:2 FTS and PFBA had the highest detection frequencies, with a detection frequency of 100%. The main PFASs were TFMS (2022: ND to 433.46 ng/L, 2023: 3.59 to 1119.56 ng/L) and 6:2 FTS (2022: 13.22 to 28.45 ng/L, 2023: 15.15 to 38.36 ng/L) (Figure 2A). These results are not entirely consistent with previous studies on aquatic PFASs in Yangtze River [33,45], which may be due to the different types of pollutants detected. Notably, TFMS was detected at a relatively high concentration, of ten reaching hundreds of ng/L, which has not been reported previously. TFMS, is a strong acid widely used as a catalyst in chemical reactions, is especially prevalent in the production of lithium-ion batteries [46,47]. The elevated TFMS concentration in surface waters are likely attributed to significant point source inputs [48], such as emissions from lithium ion battery manufacturers located in Sichuan and Chongqing [49]. As shown in Table S9, previous studies have also reported the presence of PFASs in surface waters worldwide. Overall, with the exception of TFMS (for which environmental data remain scarce), the total concentration of PFASs in the upper Yangtze River is higher than that in the Yellow River, Pearl River and Poyang Lake [14,45], and lower than that in Taihu Lake, the Thames River in the UK, and urban rivers in Europe and the United States [8,14,50]. In terms of individual PFAS, the concentration of PFOA (ND to 86.90 ng/L) and PFOS (0.36-12.12 ng/L) in this study were lower than those previously reported in Yangtze River surface water [29,34]. These differences may be due to the fact that PFOA and PFOS have been restricted in recent years [51]. PFBA, PFHxA, PFBS, and 6:2 FTS were used as substitutes for PFOA and PFOS, and their detection concentrations were higher than previously reported [14,33]. In this study, the total concentration of OPFRs in surface water ranged from 26.25 to 93.19 ng/L in 2022, and from 17.36 to 190.42 ng/L in 2023. The detection concentrations of the nine OPFRs included in the test were similar in 2022 and 2023 (Figure 2B). Furthermore, three alkyl-OPFRs (TMP, TEP, and TnBP) were found in more than 90% of the samples. Three halogenated OPFRs, TCEP, TCPP, and TDCPP, were found in more than 70% of the samples. TCPP was the main OPFR, with the concentration in the range from 10.65 to 75.57 ng/L in 2022 and in the range from ND to 86.19 ng/L in 2023, with detection frequencies > 95%. For aryl-OPEs, the detection rates of TPhP were 53% in 2022 and 62% in 2023, and the concentration ranged from ND to 5.84 ng/L in 2022 and ND to 6.96 ng/L in 2023. Generally, this study revealed that the contents of the OPEs in the water samples of the upper Yangtze River could be ranked as follows: halogenated OPFRs > alkyl-OPFRs > aryl-OPFRs. The production and application volume, as well as photodegradation may explain the different detection rates and the concentration of OPFRs [52,53]. These results were consistent with previous studies on OPFRs in Yangtze River [31,36]. The high detection rates (100%) of TMP in both 2022 and 2023 might be related to its high-water solubility (5.00×10^5 mg/L) and frequent use in China, such as solvents and catalysts in organic synthesis, pharmaceutical intermediates, plasticizers, and flame retardants [54]. The dominance of TCPP in aquatic environments could be related to its relative persistence [55]. As listed in Table S10, the levels of the OPFRs in the upper Yangtze River were comparable to that in Xiangjiang River [56], higher than that in Brazilian surface water [57], but lower than those detected in rivers and lakes in North America, Europe, and other regions of China [31,36,58–63].

Table 1. Summary of PFAS and OPFR concentration (ng/L) in water samples.

Period	Compound	Surface Water					Effluent Wastewater				
		Min. (ng/L)	Max. (ng/L)	Median (ng/L)	Mean (ng/L)	DF (%)	Min. (ng/L)	Max. (ng/L)	Median (ng/L)	Mean (ng/L)	DF (%)
	Per- and poly-fluoroalkyl substances (PFASs)										
	PFOA	ND	57.22	1.93	6.25	69.2	ND	34.30	6.42	8.75	83.3
	PFOS	ND	3.24	3.24	0.25	7.7	ND	7.57	3.60	2.52	66.7
	8:2 FTUCA	ND	4.95	1.07	1.48	76.9	ND	5.39	0.62	1.49	83.3
	6:2 FTS	13.22	28.45	19.44	20.68	100.0	14.13	34.31	21.95	23.97	100.0
	8:2 FTS	ND	5.38	0.88	1.20	76.9	ND	2.96	1.48	0.89	50.0
	PFBA	0.72	6.92	1.71	2.18	100.0	1.35	14.36	4.10	6.04	100.0
	PFBS	ND	4.95	2.69	1.22	46.2	ND	4.39	4.16	1.46	50.0
	PFHpA	ND	5.30	0.63	0.56	38.5	ND	4.25	3.82	1.73	50.0
	PFDA	ND	1.53	0.45	0.21	30.8	ND	0.65	0.29	0.20	50.0
	PFHxA	ND	14.66	5.09	4.71	84.6	ND	6.82	4.65	2.87	83.3
	PFHxS	ND	ND	ND	ND	0.0	ND	87.55	87.55	14.59	16.7
	PFNA	ND	2.86	0.70	0.58	69.2	ND	3.86	0.99	1.32	83.3
	FOSA	ND	0.12	0.12	0.01	7.7	ND	1.27	0.63	0.44	66.7
	PFPeA	ND	3.59	0.81	0.42	30.8	ND	13.03	5.62	4.23	66.7
2022	ADONA	ND	0.67	0.65	0.13	23.1	ND	0.57	0.57	0.10	16.7
	TFMS	ND	433.46	53.49	91.29	84.6	ND	726.98	39.58	168.31	83.3
	TFA	ND	0.60	0.60	0.05	7.7	ND	72.62	7.26	13.61	50.0
	HFPO-DA	ND	1.14	1.14	0.09	7.7	ND	3.03	1.11	0.77	50.0
	∑PFASs	16.07	454.68	90.27	131.43	100.0	36.41	849.82	136.97	253.27	100.0
	organophosphate flame retardants (OPFRs)										
	TPhP	ND	5.84	1.78	1.66	61.5	ND	6.96	4.47	3.27	66.7
	TPrP	ND	4.19	1.18	0.76	38.5	ND	1.52	1.14	0.38	33.3
	TMP	1.73	2.94	2.01	2.03	100.0	1.72	3.01	2.46	2.36	100.0
	TEP	ND	14.04	2.31	3.34	92.3	3.50	176.51	4.94	38.01	100.0
	TDCPP	0.50	4.33	0.88	1.58	100.0	ND	10.43	1.18	3.32	83.3
	TCEP	ND	22.14	5.22	4.15	69.2	ND	461.48	42.80	91.48	66.7
	TnBP	0.87	19.03	5.51	7.09	100.0	3.36	64.86	8.31	17.12	100.0
	TBOEP	ND	7.46	2.85	1.03	30.8	ND	240.24	3.42	40.70	50.0
	TCPP	10.65	75.57	46.27	43.98	100.0	22.38	200.86	39.59	68.18	100.0
	∑OPFRs	26.25	93.19	62.90	65.62	100.0	33.95	1007.79	92.96	264.83	100.0

Table 1. *Cont.*

Period	Compound	Surface Water Min. (ng/L)	Max. (ng/L)	Median (ng/L)	Mean (ng/L)	DF (%)	Effluent Wastewater Min. (ng/L)	Max. (ng/L)	Median (ng/L)	Mean (ng/L)	DF (%)
	Per- and poly-fluoroalkyl substances (PFASs)										
	PFOA	ND	43.19	1.61	4.90	82.4	ND	11.80	1.55	2.94	66.7
	PFOS	ND	2.20	ND	0.45	23.5	ND	5.26	ND	0.88	16.7
	8:2 FTUCA	ND	8.14	0.49	1.01	76.5	ND	0.78	0.58	0.52	83.3
	6: 2 FTS	15.15	38.36	20.11	21.37	100.0	14.17	29.11	21.74	21.58	100.0
	8:2 FTS	ND	4.57	0.84	0.89	70.6	ND	7.76	0.41	2.16	50.0
	PFBA	0.66	10.43	2.34	3.72	100.0	0.94	20.52	1.62	4.66	100.0
	PFBS	ND	20.56	ND	3.20	41.2	ND	0.70	ND	0.23	33.3
	PFHpA	ND	2.29	ND	0.56	35.3	ND	4.81	ND	0.80	16.7
	PFDA	ND	0.92	ND	0.16	29.4	ND	0.29	ND	0.05	16.7
	PFHxA	ND	13.37	3.72	4.69	76.5	ND	8.48	6.00	5.29	83.3
	PFHxS	ND	70.66	ND	4.16	5.9	ND	19.41	ND	3.24	16.7
	PFNA	ND	1.27	0.31	0.42	82.4	ND	1.82	0.19	0.41	66.7
	FOSA	ND	2.61	ND	0.20	29.4	ND	0.71	0.06	0.24	50.0
	PFPeA	ND	13.82	0.35	1.74	52.9	ND	0.98	0.20	0.39	50.0
2023	ADONA	ND	1.34	ND	0.40	52.9	ND	1.01	0.23	0.32	50.0
	TFMS	3.59	1119.56	260.44	337.15	100.0	18.61	888.25	161.47	297.23	100.0
	TFA	ND	51.44	2.79	8.36	64.7	ND	ND	ND	ND	0.0
	HFPO-DA	ND	7.56	ND	0.74	29.4	ND	0.92	ND	0.15	16.7
	∑PFASs	23.13	1231.11	302.65	394.12	100.0	47.45	926.93	220.70	341.09	100.0
	organophosphate flame retardants (OPFRs)										
	TPhP	ND	5.71	1.18	0.89	52.9	ND	13.57	9.44	3.15	33.3
	TPrP	ND	105.04	9.76	17.02	58.8	ND	189.98	2.74	32.80	66.7
	TMP	1.49	2.64	1.83	1.93	100.0	1.70	2.77	1.82	1.96	100.0
	TEP	ND	17.74	2.34	3.45	82.4	0.88	129.65	7.16	35.14	100.0
	TDCPP	ND	5.62	1.16	1.73	94.1	ND	15.53	8.96	6.21	66.7
	TCEP	ND	17.77	2.10	3.12	94.1	0.56	1221.10	19.04	227.55	100.0
	TnBP	0.83	23.93	3.96	7.01	100.0	1.59	34.42	10.26	13.90	100.0
	TBOEP	ND	8.76	1.90	1.41	47.1	ND	514.91	15.04	91.86	83.3
	TCPP	ND	86.19	40.32	39.81	88.2	23.36	572.81	59.78	136.06	100.0
	∑OPFRs	17.36	190.42	61.64	76.35	100.0	46.83	2427.21	206.76	548.61	100.0

DF: detection frequency. ND: not detected. Median: the middle concentration measurement such that half of the measurements are above it and half are below. Mean: represents the sum of all concentration measurements divided by the number of measurements (sampling points in this study). Detection frequency: refers to the proportion of measurements where the concentration exceeds the analytical method's detection limit.

Regarding effluent wastewater, the total concentration of PFASs ranged from 36.41 to 849.82 ng/L in 2022, and from 47.45 to 926.93 ng/L in 2023 (Figure 2A). The most frequently detected PFASs in effluents were 6:2 FTS (100%), PFBA (100%), and TFMS (92%). The primary PFASs detected in these samples were TFMS:ND to 726.98 ng/L in 2022, and 18.61 to 888.25 ng/L in 2023. In fact, there were no significant differences between surface water and effluent wastewater samples, either in terms of the concentration of individual compounds or total PFASs. This may be due to the strong antioxidant properties of PFASs [35]. Furthermore, the high concentration of PFASs observed in effluent wastewater suggests that WWTPs are ineffective at removing PFASs [64]. Therefore, WWTPs could be one of the main sources of PFASs in surface water [65], and lead to a high concentration in water from the upper reaches of Yangtze River. For OPFRs, the total concentration of OPFRs ranged from 33.95 to 1007.79 ng/L in 2022, and from 46.83 to 2427.21 ng/L in 2023 (Figure 2B). The most frequently detected OPFRs were TMP, TEP, TnBP, and TCPP, reaching 100%. Additionally, the highest concentration for TPrP, TEP, TCEP, TBOEP, and TCPP in effluent water were 189.98, 176.51, 1221.10, 514.91, and 572.81 ng/L, respectively. In fact, the concentration of OPFRs in effluent wastewater was much higher than that in surface water (Figure 2B). This discrepancy may be attributed to the photolysis of OPFRs after prolonged exposure to sunlight in surface waters, as well as the dilution effect of larger water bodies [31,36]. The chemical compositions of effluent water are listed in Table S11.

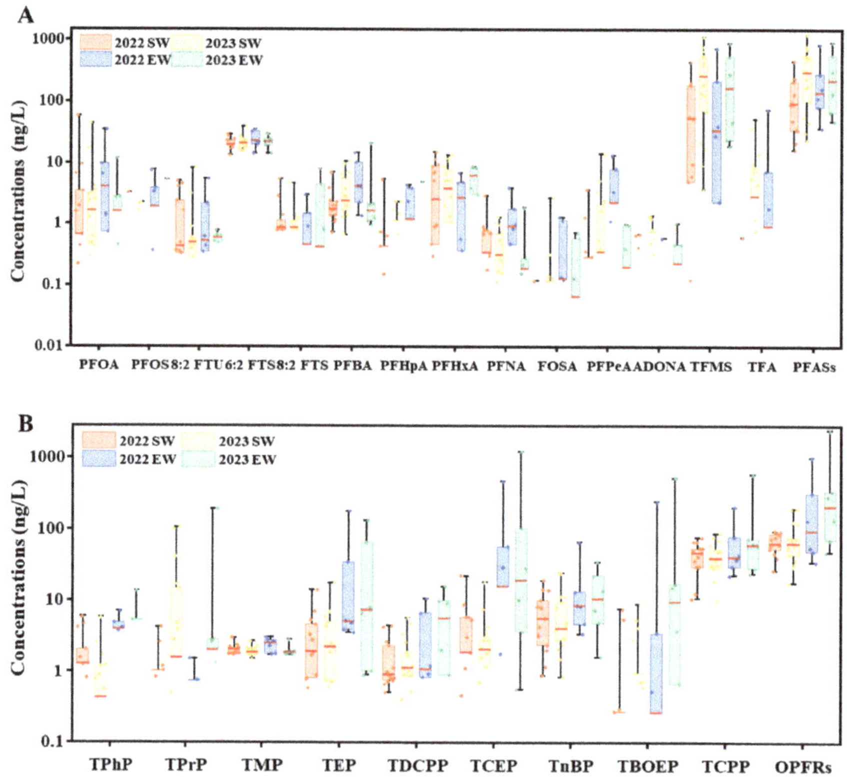

Figure 2. Concentrations of the PFASs (**A**) and OPFRs (**B**) in the upper Yangtze River in 2022 and 2023. SW: surface water, EW: effluent wastewater. Box charts (bottom to top) presented the 5th, 25th, 50th, 75th, and 95th percentile values. The absence of a box indicates that the detection rate of the compound is less than 50%.

3.2. Spatiotemporal Distribution of PFASs and OPFRs

The total mean concentration of PFASs in the main stream was higher than in the tributaries (main stream: 184.82 ng/L in 2022 and 496.84 ng/L in 2023; the tributaries: 85.46 ng/L in 2022 and 247.37 ng/L in 2023) (Figure 3A). Furthermore, the concentration of PFASs in different tributaries also varied and could be ranked as follows: Minjiang River (243.99 ng/L in 2022, 721.88 ng/L in 2023) > Tuojiang River (130.46 ng/L in 2022, 554.56 ng/L in 2023) > Jialing River (48.97 ng/L in 2022, 44.48 ng/L in 2023) > Wujiang River (33.78 ng/L in 2022, 53.88 ng/L in 2023) > Jinsha River (16.07 ng/L in 2022, 23.13 ng/L in 2023) (Figure 3A). These reflected the stability of the contribution of various tributaries in the upper reaches of Yangtze River to the main-stream PFASs in 2022 and 2023. The levels of PFASs in Jialing River, Wujiang River, and Jinsha River decreased significantly, which may relate to the sites far away from the pollution sources of urban industrial areas. From a spatial distribution perspective, the highest PFAS concentration was observed at the sampling points in the main stream in Luzhou (453.56 ng/L in 2022 and 814.18 ng/L in 2023) and Chongqing (239.06 ng/L in 2022 and 1231.11 ng/L in 2023). The degree of PFAS pollution is directly related to the population density and industrial structure of the surveyed regions [66]. These elevated concentrations may be attributed to point-source inputs from upstream fluorine chemical industrial parks located near Luzhou and Chongqing [34]. The total average concentration of OPFRs in the main stream and tributaries was similar in 2022 and 2023 (main stream: 65.62 ng/L in 2022, 76.35 ng/L in 2023 and tributaries: 66.34 in 2022, 80.92 ng/L in 2023) (Figure 3B). The concentration of OPFRs in different tributaries also varied, and their ranking in 2022 and 2023 was not entirely consistent (Figure 3B). In 2022, the tributary OPFRs levels were ranked as follows: Minjiang River (93.19 ng/L) > Jialing River (73.35 ng/L) > Tuojiang River (62.90 ng/L) > Wujiang River (52.01 ng/L) > Jinsha River (45.08 ng/L). In 2023, the ranking was Jialing River (186.95 ng/L) > Minjiang River (125.73 ng/L) > Tuojiang River (74.90 ng/L) > Jinsha River (30.11 ng/L) > Wujiang River (17.36 ng/L). High concentrations of OPFRs were detected in water bodies in Leshan and Chongqing. The population density and human activities were the important reason for the changes in the concentration of OPFRs on the spatial scale [30,55,67]. Chongqing, a municipality with a dense population and a highly developed economy, exhibited high OPFR levels. Additionally, the incomplete removal of halogenated OPFRs in wastewater by WWTPs was one of the important sources of OPFR pollution in rivers [14]. For example, high concentrations of OPFRs, especially TEP (176.51 ng/L), TCEP (1221.10 ng/L), TCPP (514.91 ng/L), and TBOEP (1221.10 ng/L), were detected in the effluents of Yibin WWTPs. The concentration of OPFRs was influenced by anthropogenic activities such as vehicle emissions, industrial effluents, and release from daily household products which may be the main sources of OPFRs. This study found that the concentration of OPFRs in urban areas and industrial areas with active human activities was higher than that in residential areas with low population density. It was reported that the release of building materials, vehicle emissions, industrial wastewater, etc. may be the main sources of OPFRs [56].

Figure 4 illustrates the changes in PFASs and OPFRs over time, highlighting significant differences in the cumulative concentration between the year of 2022 and 2023. Notably, the concentration of PFASs in 2023 was significantly higher than in 2022. NMDS analyses further confirmed the distinct differences in PFAS and OPFR concentrations between the two years (Figure S1). It is worth noting that before the sampling period in August 2023, heavy rainfall and flooding occurred in some areas of Sichuan and Chongqing, which may have affected the concentration of pollutants in the surface water. A previous study showed that higher concentration levels of PFASs in rivers during the rainy season could be attributed to the abundant precipitation and runoff in urban rivers [68].

Figure 5 displays the composition profiles of individual PFASs and OPFRs. Among the PFASs, TFMS had with the largest contribution, ranging from 0–96% in 2022 and 16–97% in 2023. Furthermore, at some sampling points (D6, D7 in Luzhou, D8, D9 in Chongqing), the composition of PFASs also differed in 2022 and 2023 (Figure 5A). The PFASs detected in the waters of the upper Yangtze River were predominantly ultra-short-chain PFASs (e.g., TFMS, TFA) and FTS compounds (e.g., 6:2 FTS). This composition differed from previous reports on Yangtze River [33,34,45]. Recent studies have suggested that ultra-short-chain PFASs are widely prevalent in aquatic environments globally and are often found at higher concentrations than long-chain compounds [48], which aligns with the findings of this study. Halogenated OPFRs were the OPFRs that contribute the most to the surface water in the upper reaches of Yangtze River. TCPP was the dominant compound, contributing 20–90% in 2022 and 0–84% in 2023. Notably, there were significant differences in the composition of alkyl OPFRs in surface water in 2022 and 2023, especially TPrP (0–5% in 2022, 0–62% in 2023) and TnBP (1–30% in 2022, 2–38% in 2023). This may be partly attributed to the photolysis of aryl OPFRs due to the relatively low air temperature and

UV intensity during August 2023, when heavy rains were experienced. These results are consistent with prior studies on OPFRs in the aquatic environments of Yangtze River [36,37].

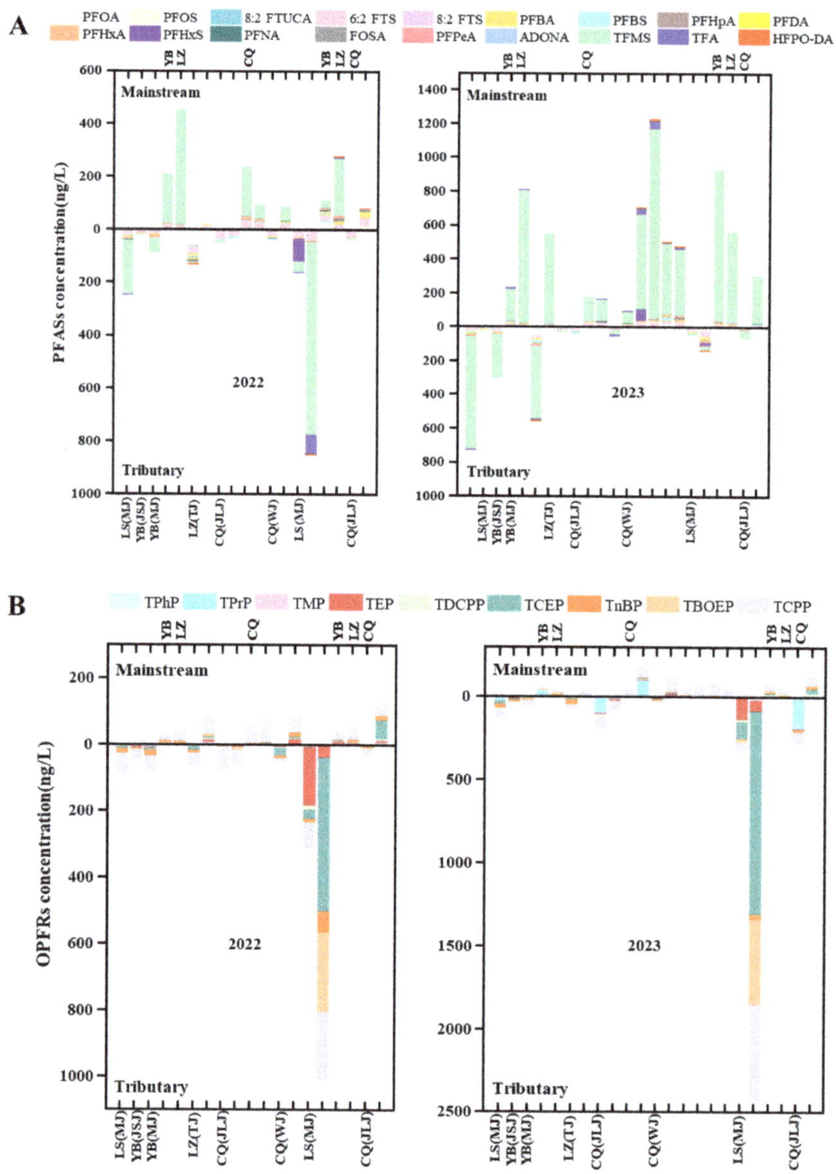

Figure 3. Total PFAS (**A**) and OPFR (**B**) concentrations (ng/L) in samples of Yangtze River in 2022 and 2023. For each figure, the upright bar refers to the main stream and the inverted bar refers to the tributaries.

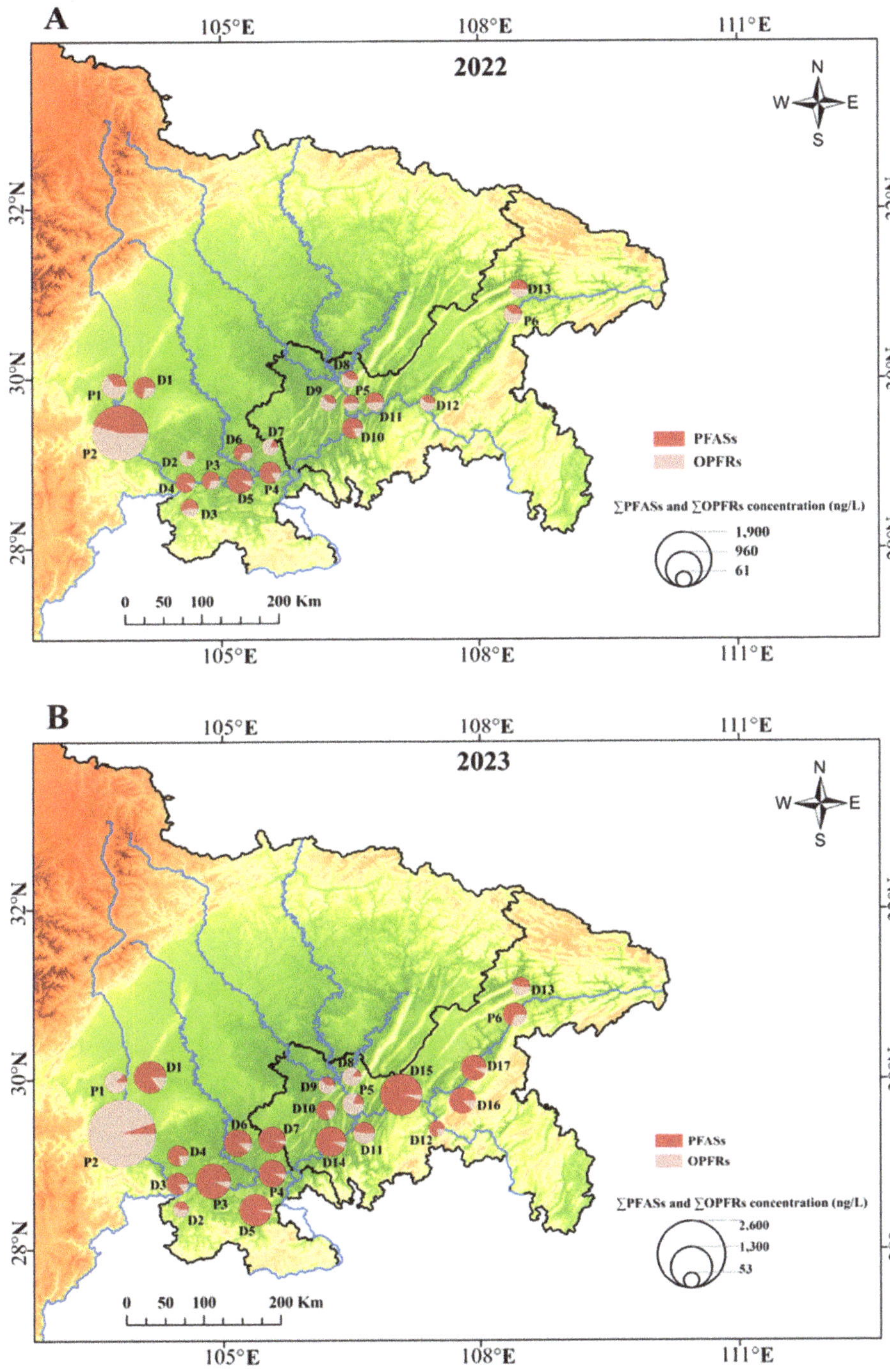

Figure 4. Cumulative concentrations of PFASs and OPFRs in the samples collected during 2022 (**A**) and 2023 (**B**) in the upper Yangtze River.

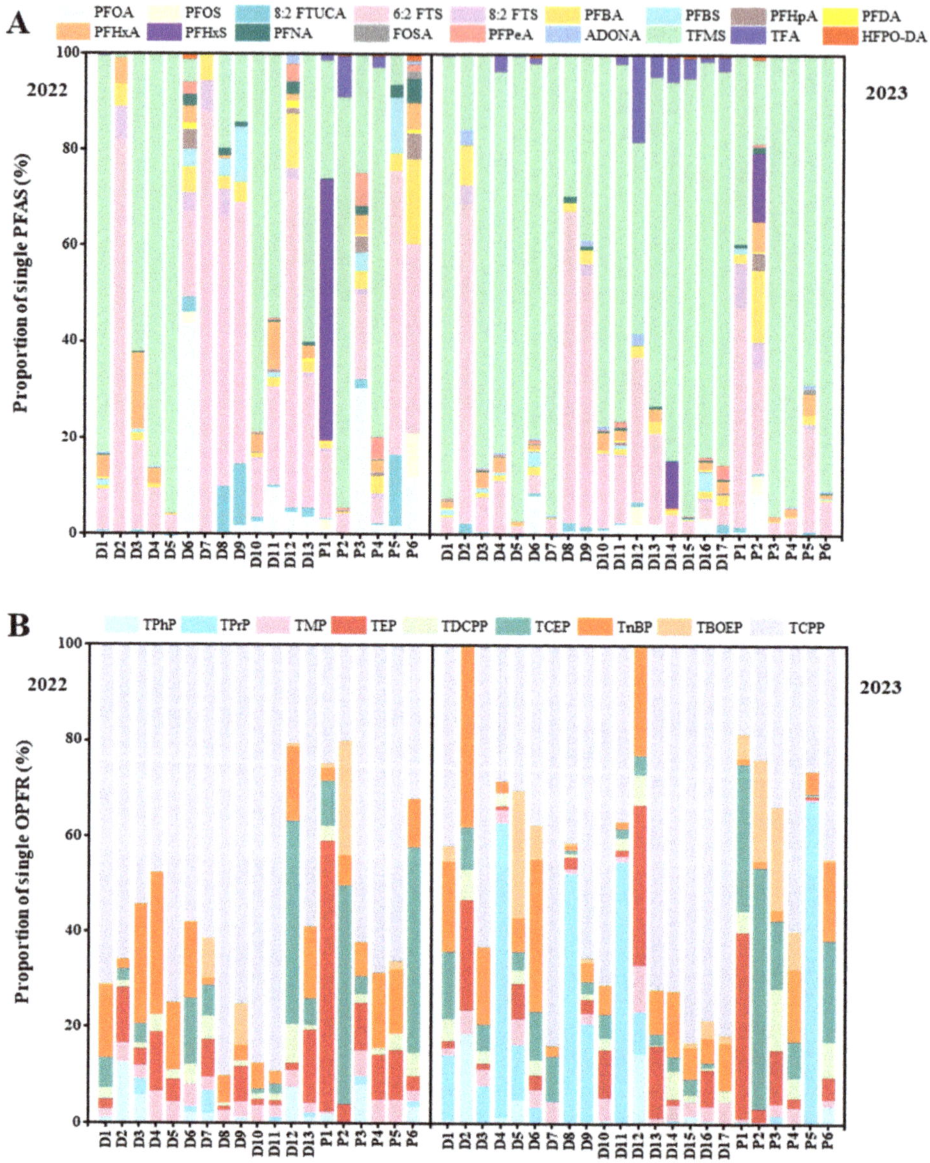

Figure 5. Relative contributions of individual composition to the total PFASs (**A**) and OPFRs (**B**) in water samples from the upper Yangtze River.

3.3. Possible Source Analysis

Spearman analysis was conducted to explore the correlation between the concentration of PFASs and OPFRs [33,36], with the results shown in Tables S12 and S13 and Figure S2. Direct sources of PFASs include industrial production, transportation, use and disposal, and indirect sources are the degradation of precursor compounds [69]. Significant positive correlations were observed among certain PFASs in Yangtze River water samples ($p < 0.05$), suggesting the presence of common contamination sources. As shown in Table S12, strong correlation was identified between long-chain PFCAs (PFOA, PFNA, PFDA) and short-chain PFCAs (PFBA, PFPeA, PFHpA) (r > 0.60; $p < 0.001$). Additionally, HFPO-DA was strongly correlated with PFOS, PFBA, and PFPeA (r > 0.60; $p < 0.001$), further supporting the idea of shared contamination sources. There are several industrial parks related to the fluorine industry along Yangtze River, such as the fluorochemical industrial parks along the upper stream near Luzhou and Chongqing [34], which may be an important source of the PFOS, PFBA, PFPeA, and long-chain and short-chain PFCAs. Additionally, due to backward sewage treatment technology, industrial wastewater containing PFASs could easily enter Yangtze River and its drainage system. It was reported that large-scale fluorine chemical industrial parks may cause large-scale surface water pollution [70]. However, no significant correlations were detected between other PFASs. PFOA, PFNA, and PFDA are key precursors and processing aids in fluoropolymer production [71,72], while PFBA, PFPeA, and PFHpA are primarily used in the manufacture of products such as carpets, gloves, leather, ski wax, and outdoor textiles [73]. These findings suggested that the industries involved in producing precursors, processing aids, leather and textile products could be potential sources of these PFCAs. Municipal wastewater often contains high concentrations of PFASs due to the daily use of products containing PFASs and inefficiencies in wastewater treatment [74]. Serious PFAS contamination has been found in the inlet and outlet water of WWTPs in several cities along Yangtze River [34]. Given that the megacity of Chongqing is located along Yangtze River, there is a significant risk of PFASs in urban sewage being discharged into the upper Yangtze River.

No significant correlations were found between OPFRs in water bodies, suggesting that these compounds OPFRs may originate from diverse sources of contamination. Halogenated alkyl OPFRs (TCEP, TCPP, and TDCPP) are commonly used as flame retardants in furniture, electronic products, and textiles [55]. Meanwhile, alkyl OPFRs (TMP, TEP, TPrP, TnBP, and TBOEP) and aryl OPFRs (TPhP) are widely utilized as plasticizers, lubricants, and brighteners in textiles, building materials, and electronic products [75]. Therefore, the sources of OPFRs in the surface water of the upper Yangtze River are mainly caused by anthropogenic activities, including the use of building materials in daily life, the release of chemicals from household products, and industrial wastewater discharge. Additionally, atmospheric transport and sedimentation were the important sources of OPFRs (such as TCEP, TCPP, TBOEP, and TnBP) in surface water [76,77].

3.4. Estimated Annual Mass Loadings of PFASs and OPFRs in the Upper Yangtze River

Figure S3 and Table S14 show the estimated annual inflows of PFASs and OPFRs for the five major tributaries and the main stream. The flux (ton/year) of PFASs in the Yangtze River main stream was 50.93 in 2022 and 139.16 in 2023; in the Jinsha River, it was 1.73 in 2022 and 2.79 in 2023; in the Min River, it was 9.88 in 2022 and 34.08 in 2023; in the Tuo River, it was 1.27 in 2022 and 5.70 in 2023; in the Jialing River, it was 1.97 in 2022 and 2.10 in 2023; in the Wu River, it was 1.20 in 2022 and 1.64 in 2023. Therefore, the tributary that contributed the most to PFASs in the mainstream is the Minjiang River (Figure S3). The calculated annual emission loadings of TFMS, 6:2 FTS, PFOA, PFOS, PFBS, PFHxA and the total PFASs in the upper Yangtze River in 2022 were 41.72 tons, 5.84 tons, 0.89 tons, 0.0 tons, 0.04, 1.31 tons, and 50.93 tons, respectively. And these values of TFMS, 6:2 FTS, PFOA, PFOS, PFBS, PFHxA and the total PFASs in 2023 were 121.04 tons, 6.31 tons, 0.93 tons, 0.09 tons, 0.67 tons, 1.47 tons, and 139.16 tons, respectively. Except for TFMS, these values were lower than previously estimated annual emission fluxes from Yangtze River to the East China Sea (PFOA: 6.8 tons; PFOS: 0.88 tons; PFHxA: 2.2 tons; PFBS: 8.2 tons, PFASs: 20.7 tons) [33], suggesting that the middle and lower Yangtze River contributed the majority of PFASs. Although the PFAS concentration in the upper Yangtze River (max. values for PFOA and PFOS were 57.22 and 3.24 ng/L) were much lower than those in the majority rivers in the world (in Japan, PFOA and PFOS were 107 ng/L and 143 ng/L; in Florida, USA, PFOA and PFOS were 81 and 1135 ng/L) [78,79], the mass loading of PFASs in the upper Yangtze River were still higher than those in other rivers due to its huge annual water flow (annual mass loading of PFOA and PFOA in Japan were 0.045 tons and 0.060 tons) [79]. Considering the persistence and bioaccumulation, PFASs in the upper Yangtze River would be transported to the middle and lower reaches, and the East China Sea, and eventually enter the ocean, which could have significant environmental impacts.

For OPFRs, the flux (ton/year) in the main stream was 17.85 in 2022 and 20.49 in 2023; in the Jinsha River, it was 4.84 in 2022 and 3.64 in 2023; in the Min River, it was 5.00 in 2022 and 6.12 in 2023; in the Tuo River, it was 0.61 in 2022 and 0.77 in 2023; in the Jialing River, it was 3.19 in 2022 and 7.23 in 2023; in the Wu River, it was 1.84 in 2022 and 0.53 in 2023. It is obvious that the Minjiang River and Jialing River contribute significantly to the annual mass loadings of mainstream OPFRs (Figure S3). The calculated annual emission loadings of total OPFRs in the upper Yangtze River were lower than that from Yangtze River to the East China Sea (97.0 tons) [36]. Therefore, the middle and lower reaches may be the main contributors to OPFRs in Yangtze River, which may be related to the high population density and developed industry in the middle and lower reaches [80].

3.5. Ecological Risks Assessment

The ecological risks of individual and total RQs of PFASs and OPFRs in water samples were evaluated based on previously reported PNECs [81–94], with the results presented in Table S15 and Figure 6. Overall, the RQ_{STU} (summed risk quotient) values for both PFASs and OPFRs across the stream sampling points in all cities ranged between 0.01 and 0.1 (Figure 6A), indicating a low ecological risk in the upper Yangtze River. However, the RQ_{STU} of OPFRs at some sampling points in Chongqing suggested a medium risk (RQ: 0.1–1). Notably, a high risk (RQ > 1) of OPFRs was detected near the wastewater effluent in Leshan. Given the potential adverse effects of PFASs and OPFRs in surface waters, further studies on the chronic toxicity of these pollutants in aquatic organisms are essential for more comprehensive risk assessments, particularly for OPFRs.

As shown in Figure 6B,C, all PFASs were unlikely to have serious adverse effects on aquatic organisms (RQs < 0.1). Although PFASs in the upper Yangtze River appear to be relatively risk-free at present, their presence should not be ignored because of the persistence, accumulation, and migration. Moreover, for the main PFASs detected, TFMS and 6:2FTS, due to the lack of toxicity studies on ultra-short-chain PFASs and FTS-related compounds, attention should be paid to their possible environmental impacts. For OPFRs, TPhP, TCEP, and TBOEP showed a low risk (0.01 < RQs < 0.1) to three typical organisms at some points in 2022 (Figure 6B). However, TCEP presented a medium risk (0.1 < RQs < 1) for Algae at some sampling points (Figure 6C). The relatively higher ecological risks of TPhP, TCEP, and TBOEP were partly due to their high acute toxicity to aquatic organisms and relatively low PNEC values (Table S6). Compared with previous studies, the ecological risk of OPFRs in the upper Yangtze River surface water was lower in this study [36,37]. Considering the long-term exposure, cumulative adverse effects, and bioaccumulation of OPFRs in aquatic environments [17,21], more studies on the chronic toxicity of OPFRs in aquatic organisms are needed for risk assessment, especially for TPhP, TCEP, and TBOEP.

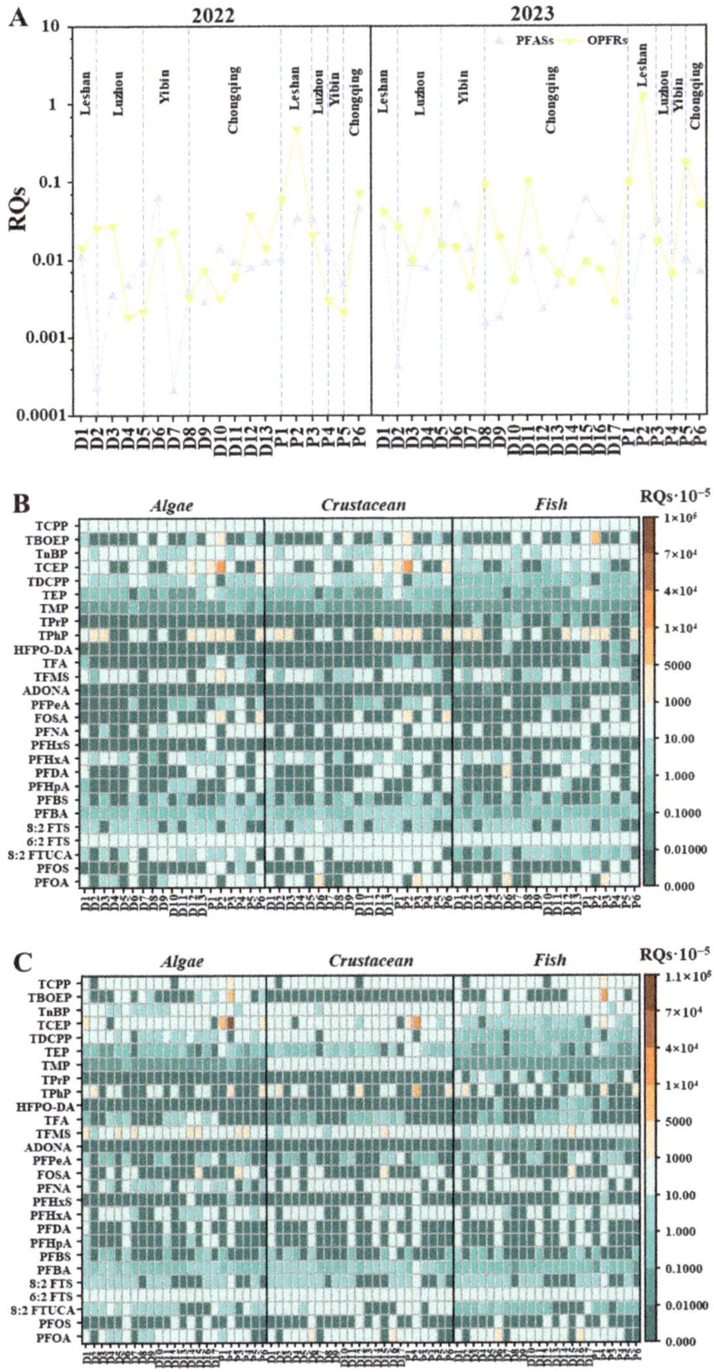

Figure 6. Risk quotients of PFASs and OPFRs. (**A**) The total risk quotients (RQ_{STU}). Ecological risks related to individual compounds in 2022 (**B**) and 2023 (**C**) for algae, crustaceans, and fish in each water sample.

3.6. Health Risk Assessment

Health risks from chemicals in aquatic environments are often linked to the human ingestion of drinking water [17,95]. Since some of the sampling points in this study were near water sources, drinking water intake was chosen as the basis for the health risk assessment. The results of hazard indices (His) and carcinogenic risks (CRs) for health-related PFASs and OPFRs are presented in Figure 7 and Table S16. For both individual compounds and the total concentration of PFASs (∑PFASs) and OPFRs (∑OPFRs), the non-carcinogenic (HI < 0.2) risk values were below theoretical risk (HI < 1) thresholds, suggesting that exposure to PFASs and OPFRs via surface water is likely safe for local residents. In general, the non-carcinogenic (non-CR) risk of PFASs was higher than that of OPFRs (Figure 7A). Notably, the non-carcinogenic risk of PFASs in Luzhou (Tuo River) and Chongqing (mainstream) was higher than that at other sampling points (Figure 7A). Regarding the age groups, the non-carcinogenic risk for PFASs followed the order: adult (2.50×10^{-1} in 2022 and 4.91×10^{-1} in 2023) > child (1.76×10^{-1} in 2022 and 1.60×10^{-1} in 2023) > teenage (1.13×10^{-1} in 2022 and 1.03×10^{-1} in 2023) > infant (6.70×10^{-2} in 2022 and 1.45×10^{-1} in 2023) (Table S16). The non-carcinogenic risk for OPFRs followed: child (1.48×10^{-3} in 2022 and 2.97×10^{-3} in 2023) > infant (1.34×10^{-3} in 2022 and 2.69×10^{-3} in 2023) > teenage (9.50×10^{-4} in 2022 and 1.90×10^{-3} in 2023) > adult (9.67×10^{-4} in 2022 and 1.43×10^{-3} in 2023) (Table S16). The non-CR risk level of PFASs in this study was higher than that of Chongqing surface water [37], while the non-CR risk level of OPFRs was lower than that of previous studies on Yangtze River surface water [32].

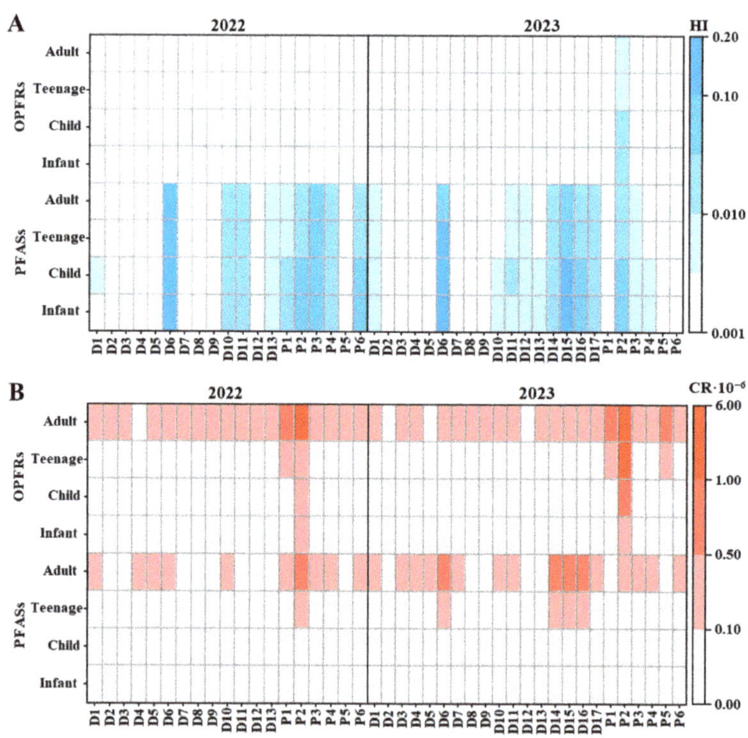

Figure 7. Health risk assessment. (**A**) Non-carcinogenicity (HI) and (**B**) carcinogenicity risk (CR) in each water sample.

For carcinogenic risk, the CR values of PFASs at all sampling points were significantly lower than 10^{-6}, indicating that the CR values associated with these pollutants within are acceptable levels. However, OPFRs at the Leshan (Min River) sampling points were estimated to pose a potential cancer risk to adults (10^{-6} to 10^{-4}) (Figure 7B). The CR order for different age groups for PFASs was as follows: adult (9.96×10^{-8} in 2022 and 1.34×10^{-7} in 2023) > teenage (9.41×10^{-8} in 2022 and 1.26×10^{-7} in 2023) > child (5.18×10^{-8} in 2022 and 6.95×10^{-8} in 2023) > infant (2.58×10^{-8} in 2022 and 3.91×10^{-8} in 2023) (Table S16). The carcinogenic risk for OPFRs was as follows: adult (5.28×10^{-7} in 2022 and 1.66×10^{-7} in 2023) > teenage (1.10×10^{-7} in 2022 and 1.57×10^{-7} in 2023) > child (5.72×10^{-8} in 2022 and 8.66×10^{-8} in 2023) > infant (2.53×10^{-8} in 2022 and 3.14×10^{-8} in 2023) (Table S16). The health risks of OPFRs in these results were consistent with those in the middle and lower reaches of Yangtze River [30,31]. In order to fully elucidate the future exposure risks of PFASs and OPFRs to humans, more target compounds and their spatiotemporal variation characteristics need to be identified and studied. In addition, the available parameters (e.g., RfD and SF) of many PFASs and OPFRs are still unknown, which will lead to an underestimation of the associated risks.

4. Conclusions

Generally, the total concentration of PFASs in the upper Yangtze River ranged from 16.07 to 1231.11 ng/L, and OPFRs ranged from 17.36 to 190.42 ng/L. Ultra-short-chain PFASs (TFMS: ND to 1119.56 ng/L) were the most prevalent PFASs in the waters of the upper reaches of Yangtze River, while 6:2FTS and PFBA were prevalent in all samples. Halogenated alkyl OPFRs (TCPP and TCEP) were commonly found in the samples, especially in the effluent wastewater, with the highest concentrations reaching 1221.10 and 572.81 ng/L. Human activities and WWTP wastewater may be the main sources of PFASs and OPFRs in the samples. Furthermore, seasonal and hydrological factors further influenced the distribution patterns, with higher pollutant loads observed during the rainy season, likely due to increased runoff and urban wastewater overflow. In the upper Yangtze River, the concentration of PFASs in the main stream was higher than that in the tributaries, while there was no significant difference in OPFRs between the main stream and tributaries. The annual mass load calculations demonstrated that Minjiang River and Jialing River contribute significantly to PFASs and OPFRs in the main stream. The ecological risk assessments indicated that PFASs and OPFRs at most sampling sites posed low-to-moderate risks to aquatic organisms, including algae, crustaceans, and fish. However, medium ecological risks were observed for specific OPFRs, such as TCPP and TCEP, particularly near wastewater discharge points in regions with intensive industrial activities. Human health risk assessments revealed that non-carcinogenic risks from PFASs and OPFRs were generally below acceptable thresholds for all age groups. However, the carcinogenic risk of the adults associated with OPFRs was elevated at certain locations, particularly in Leshan City along the Min River. Therefore, it is crucial to enhance the supervision of PFASs and OPFRs in surface water while emphasizing the need for improved wastewater treatment technologies to mitigate the environmental and health impacts of these pollutants. Future research should focus on understanding the chronic toxicity and bioaccumulation of ultra-short-chain PFASs and halogenated alkyl OPFRs to support accurate ecological risk assessments and management policies.

Supplementary Materials: The following supporting information can be downloaded at: https://www.mdpi.com/article/10.3390/toxics13020116/s1, Text S1. Chemicals; Text S2. Sample preparation; Table S1. Summary of physico-chemical information of PAFAs and OPFRs in this study; Table S2. Sampling point information and water quality parameters; Table S3. The instrument details and gradient elution conditions; Table S4. LODs and LOQs of PFASs and OPFRs; Table S5. The runoff (108 m^3) of the upper Yangtze River and its tributaries; Table S6. Toxicity values and risk quotients (RQs) for the aquatic organisms; Table S7. Summary of the available RfD (ng/kg/day) and SF (1/(ng/kg/day)); Table S8. Parameters used in human exposure estimation and health risk assessments; Table S9. Summary of the median and range (ng/L) of PFASs in water reported in the published literature; Table S10. Summary of the median and range (ng/L) of OPFRs in water reported in the published literature; Table S11. The chemical composition (organic contaminants) of the main industrial effluent wastewater; Table S12. Spearman rank order correlation coefficients of PFASs in water samples; Table S13. Spearman rank order correlation coefficients of OPFRs in water samples; Table S14. Annual emission flux (F: tons/year) of PFASs and OPFRs discharged into the upper Yangtze River; Table S15. Risk quotients (RQs) of PFASs and OPFRs for the aquatic organisms in the upper Yangtze River; Table S16. Non-carcinogenic risk (HI values) and carcinogenic risk (CR) of PFASs and OPFRs in the upper Yangtze River; Figure S1. NMDS analyses of PFAS (A) and OPFR (B) concentrations in water samples in 2022 and 2023; Figure S2. The calculated correlation between each concentration of PFASs (A) and OPFRs (B) based on non-parametric measure of the Spearman correlation; Figure S3. Annual emission flux (F: tons/year) of PFASs and OPFRs.

Author Contributions: Conceptualization, W.S. and C.W.; methodology, W.S., Y.B., C.W. and Z.F.; investigation, W.S., Y.L. and C.W.; formal analysis, C.W., Y.B., Y.Z. and Z.F.; visualization, W.S.; data curation, W.S. and C.W.; experimental design, F.W. and Y.Z.; writing—original draft preparation, W.S.; writing—review and editing W.S., C.W. and F.W.; supervision, C.W., Y.B., Y.Z. and F.W.; project administration, C.W. and Z.F.; funding acquisition, C.W. and Z.F. All authors have read and agreed to the published version of the manuscript.

Funding: This research was funded by the National Key Research and Development Program of China (2021YFC3200801).

Institutional Review Board Statement: Not applicable.

Informed Consent Statement: Not applicable.

Data Availability Statement: Data are available in the article.

Conflicts of Interest: The authors declare no conflicts of interest.

References

1. Halden, R.U. Epistemology of contaminants of emerging concern and literature meta-analysis. *J. Hazard. Mater.* **2015**, *282*, 2–9. [CrossRef] [PubMed]
2. Fairbairn, D.J.; Arnold, W.A.; Barber, B.L.; Kaufenberg, E.F.; Koskinen, W.C.; Novak, P.J.; Rice, P.J.; Swackhamer, D.L. Contaminants of Emerging Concern: Mass Balance and Comparison of Wastewater Effluent and Upstream Sources in a Mixed-Use Watershed. *Environ. Sci. Technol.* **2016**, *50*, 36–45. [CrossRef]
3. Kim, U.-J.; Kannan, K. Occurrence and Distribution of Organophosphate Flame Retardants/Plasticizers in Surface Waters, Tap Water, and Rainwater: Implications for Human Exposure. *Environ. Sci. Technol.* **2018**, *52*, 5625–5633. [CrossRef] [PubMed]
4. Li, T.; Chen, Y.; Wang, Y.; Tan, Y.; Jiang, C.; Yang, Y.; Zhang, Z. Occurrence, source apportionment and risk assessment of perfluorinated compounds in sediments from the longest river in Asia. *J. Hazard. Mater.* **2024**, *467*, 133608. [CrossRef]
5. Liu, T.; Qian, X.; Wang, S.; Wang, H.; Wei, S.; Chen, H. Occurrence and transport of perfluoroalkyl acids (PFAAs) in a Yangtze River water diversion project during water diversion and flooding. *Water Res.* **2021**, *205*, 117662. [CrossRef]
6. Xu, K.; Huang, J.; Zhang, Y.; Wu, X.; Cai, D.; Hu, G.; Li, Y.; Ni, Z.; Lin, Q.; Wang, S.; et al. Crop Contamination and Human Exposure to Per- and Polyfluoroalkyl Substances around a Fluorochemical Industrial Park in China. *Toxics* **2024**, *12*, 269. [CrossRef]
7. Medina, H.; Farmer, C. Current Challenges in Monitoring Low Contaminant Levels of Per- and Polyfluoroalkyl Substances in Water Matrices in the Field. *Toxics* **2024**, *12*, 610. [CrossRef]

8. Nakayama, S.F.; Yoshikane, M.; Onoda, Y.; Nishihama, Y.; Iwai-Shimada, M.; Takagi, M.; Kobayashi, Y.; Isobe, T. Worldwide trends in tracing poly- and perfluoroalkyl substances (PFAS) in the environment. *TrAC Trends Anal. Chem.* **2019**, *121*, 115410. [CrossRef]
9. Stecconi, T.; Stramenga, A.; Tavoloni, T.; Bacchiocchi, S.; Ciriaci, M.; Griffoni, F.; Palombo, P.; Sagratini, G.; Siracusa, M.; Piersanti, A. Exploring Perfluoroalkyl Substances (PFASs) in Aquatic Fauna of Lake Trasimeno (Italy): Insights from a Low-Anthropized Area. *Toxics* **2024**, *12*, 196. [CrossRef] [PubMed]
10. Da, S.; Wang, J. Occurrence, Bioaccumulation, and Risk Assessment of Organophosphate Esters in Rivers Receiving Different Effluents. *Toxics* **2024**, *12*, 612. [CrossRef]
11. Shi, Y.; Gao, L.; Li, W.; Wang, Y.; Liu, J.; Cai, Y. Occurrence, distribution and seasonal variation of organophosphate flame retardants and plasticizers in urban surface water in Beijing, China. *Environ. Pollut.* **2016**, *209*, 1–10. [CrossRef]
12. Rodgers, T.F.M.; Giang, A.; Diamond, M.L.; Gillies, E.; Saini, A. Emissions and fate of organophosphate esters in outdoor urban environments. *Nat. Commun.* **2023**, *14*, 1175. [CrossRef]
13. Wang, J.; Lin, J.; Zhang, X.; Zeng, Q.; Zhu, Z.; Zhao, S.; Cao, D.; Zhu, M. Organophosphate Esters and Polybrominated Diphenyl Ethers in Vehicle Dust: Concentrations, Sources, and Health Risk Assessment. *Toxics* **2024**, *12*, 806. [CrossRef]
14. Pan, Y.; Zhang, H.; Cui, Q.; Sheng, N.; Yeung, L.W.Y.; Sun, Y.; Guo, Y.; Dai, J. Worldwide Distribution of Novel Perfluoroether Carboxylic and Sulfonic Acids in Surface Water. *Environ. Sci. Technol.* **2018**, *52*, 7621–7629. [CrossRef]
15. Wang, Y.; Hou, M.; Zhang, Q.; Wu, X.; Zhao, H.; Xie, Q.; Chen, J. Organophosphorus Flame Retardants and Plasticizers in Building and Decoration Materials and Their Potential Burdens in Newly Decorated Houses in China. *Environ. Sci. Technol.* **2017**, *51*, 10991–10999. [CrossRef]
16. Zhang, Y.; Lv, Z.; Yu, X.-Y.; Zhang, Y.; Zhu, L. Integration of Nontarget Screening and QSPR Models to Identify Novel Organophosphate Esters of High Priority in Aquatic Environment. *Environ. Sci. Technol.* **2024**, *58*, 14506–14517. [CrossRef] [PubMed]
17. Wang, X.; Zhu, Q.; Yan, X.; Wang, Y.; Liao, C.; Jiang, G. A review of organophosphate flame retardants and plasticizers in the environment: Analysis, occurrence and risk assessment. *Sci. Total Environ.* **2020**, *731*, 139071. [CrossRef] [PubMed]
18. Jensen, C.R.; Genereux, D.P.; Solomon, D.K.; Knappe, D.R.U.; Gilmore, T.E. Forecasting and Hindcasting PFAS Concentrations in Groundwater Discharging to Streams near a PFAS Production Facility. *Environ. Sci. Technol.* **2024**, *58*, 17926–17936. [CrossRef] [PubMed]
19. Sadia, M.; Nollen, I.; Helmus, R.; ter Laak, T.L.; Béen, F.; Praetorius, A.; van Wezel, A.P. Occurrence, Fate, and Related Health Risks of PFAS in Raw and Produced Drinking Water. *Environ. Sci. Technol.* **2023**, *57*, 3062–3074. [CrossRef] [PubMed]
20. Wang, L.; Yang, T.; Liu, X.; Liu, J.; Liu, W. Critical Evaluation and Meta-Analysis of Ecotoxicological Data on Per- and Polyfluoroalkyl Substances (PFAS) in Freshwater Species. *Environ. Sci. Technol.* **2024**, *58*, 17555–17566. [CrossRef] [PubMed]
21. Yao, C.; Yang, H.; Li, Y. A review on organophosphate flame retardants in the environment: Occurrence, accumulation, metabolism and toxicity. *Sci. Total Environ.* **2021**, *795*, 148837. [CrossRef]
22. Pickard, H.M.; Ruyle, B.J.; Haque, F.; Logan, J.M.; LeBlanc, D.R.; Vojta, S.; Sunderland, E.M. Characterizing the Areal Extent of PFAS Contamination in Fish Species Downgradient of AFFF Source Zones. *Environ. Sci. Technol.* **2024**, *58*, 19440–19453. [CrossRef]
23. Podder, A.; Sadmani, A.H.M.A.; Reinhart, D.; Chang, N.-B.; Goel, R. Per and poly-fluoroalkyl substances (PFAS) as a contaminant of emerging concern in surface water: A transboundary review of their occurrences and toxicity effects. *J. Hazard. Mater.* **2021**, *419*, 126361. [CrossRef] [PubMed]
24. Hong, X.; Yuan, L.; Zhao, X.; Shan, Y.; Qin, T.; Li, J.; Zha, J. Embryonic Exposure to Organophosphate Flame Retardants (OPFRs) Differentially Induces Cardiotoxicity in Rare Minnow (*Gobiocypris rarus*). *Environ. Sci. Technol.* **2024**, *58*, 13648–13657. [CrossRef] [PubMed]
25. Li, X.; Li, N.; Rao, K.; Huang, Q.; Ma, M. In Vitro Immunotoxicity of Organophosphate Flame Retardants in Human THP-1-Derived Macrophages. *Environ. Sci. Technol.* **2020**, *54*, 8900–8908. [CrossRef]
26. Liu, Y.; Gong, S.; Ye, L.; Li, J.; Liu, C.; Chen, D.; Fang, M.; Letcher, R.J.; Su, G. Organophosphate (OP) diesters and a review of sources, chemical properties, environmental occurrence, adverse effects, and future directions. *Environ. Int.* **2021**, *155*, 106691. [CrossRef] [PubMed]
27. Park, H.; Choo, G.; Kim, H.; Oh, J.-E. Evaluation of the current contamination status of PFASs and OPFRs in South Korean tap water associated with its origin. *Sci. Total Environ.* **2018**, *634*, 1505–1512. [CrossRef] [PubMed]
28. Liang, H.; Zhang, D.; Wang, W.; Yu, S.; Nimai, S. Evaluating future water security in the upper Yangtze River Basin under a changing environment. *Sci. Total Environ.* **2023**, *889*, 164101. [CrossRef]
29. Wang, A.; Yang, D.; Tang, L. Spatiotemporal variation in nitrogen loads and their impacts on river water quality in the upper Yangtze River basin. *J. Hydrol.* **2020**, *590*, 125487. [CrossRef]
30. Lian, M.; Lin, C.; Li, Y.; Hao, X.; Wang, A.; He, M.; Liu, X.; Ouyang, W. Distribution, partitioning, and health risk assessment of organophosphate esters in a major tributary of middle Yangtze River using Monte Carlo simulation. *Water Res.* **2022**, *219*, 118559. [CrossRef]

31. Xing, L.; Tao, M.; Zhang, Q.; Kong, M.; Sun, J.; Jia, S.; Liu, C.-H. Occurrence, spatial distribution and risk assessment of organophosphate esters in surface water from the lower Yangtze River Basin. *Sci. Total Environ.* **2020**, *734*, 139380. [CrossRef]
32. Yu, L.; Liu, X.; Hua, Z. Occurrence, distribution, and risk assessment of perfluoroalkyl acids in drinking water sources from the lower Yangtze River. *Chemosphere* **2022**, *287*, 132064. [CrossRef]
33. Pan, C.; Ying, G.; Zhao, J.; Liu, Y.; Jiang, Y.; Zhang, Q. Spatiotemporal distribution and mass loadings of perfluoroalkyl substances in the Yangtze River of China. *Sci. Total Environ.* **2014**, *493*, 580–587. [CrossRef]
34. Li, J.; Gao, Y.; Xu, N.; Li, B.; An, R.; Sun, W.; Borthwick, A.G.L.; Ni, J. Perfluoroalkyl substances in the Yangtze River: Changing exposure and its implications after operation of the Three Gorges Dam. *Water Res.* **2020**, *182*, 115933. [CrossRef] [PubMed]
35. Lorenzo, M.; Campo, J.; Morales Suárez-Varela, M.; Picó, Y. Occurrence, distribution and behavior of emerging persistent organic pollutants (POPs) in a Mediterranean wetland protected area. *Sci. Total Environ.* **2019**, *646*, 1009–1020. [CrossRef] [PubMed]
36. Li, S.; Wan, Y.; Wang, Y.; He, Z.; Xu, S.; Xia, W. Occurrence, spatial variation, seasonal difference, and ecological risk assessment of organophosphate esters in the Yangtze River, China: From the upper to lower reaches. *Sci. Total Environ.* **2022**, *851*, 158021. [CrossRef] [PubMed]
37. Zhao, J.; Guo, C.; Yang, Q.; Liu, W.; Zhang, H.; Luo, Y.; Zhang, Y.; Wang, L.; Chen, C.; Xu, J. Comprehensive monitoring and prioritizing for contaminants of emerging concern in the Upper Yangtze River, China: An integrated approach. *J. Hazard. Mater.* **2024**, *480*, 135835. [CrossRef] [PubMed]
38. Rex, K.R.; Chakraborty, P. Legacy and new chlorinated persistent organic pollutants in the rivers of south India: Occurrences, sources, variations before and after the outbreak of the COVID-19 pandemic. *J. Hazard. Mater.* **2022**, *437*, 129262. [CrossRef] [PubMed]
39. Liu, N.; Jin, X.; Feng, C.; Wang, Z.; Wu, F.; Johnson, A.C.; Xiao, H.; Hollert, H.; Giesy, J.P. Ecological risk assessment of fifty pharmaceuticals and personal care products (PPCPs) in Chinese surface waters: A proposed multiple-level system. *Environ. Int.* **2020**, *136*, 105454. [CrossRef]
40. Backhaus, T.; Faust, M. Predictive Environmental Risk Assessment of Chemical Mixtures: A Conceptual Framework. *Environ. Sci. Technol.* **2012**, *46*, 2564–2573. [CrossRef]
41. Liu, J.; Lu, G.; Yang, H.; Dang, T.; Yan, Z. Ecological impact assessment of 110 micropollutants in the Yarlung Tsangpo River on the Tibetan Plateau. *J. Environ. Manag.* **2020**, *262*, 110291. [CrossRef]
42. Legay, C.; Rodriguez, M.J.; Sadiq, R.; Sérodes, J.B.; Levallois, P.; Proulx, F. Spatial variations of human health risk associated with exposure to chlorination by-products occurring in drinking water. *J. Environ. Manag.* **2011**, *92*, 892–901. [CrossRef] [PubMed]
43. Li, J.; Zhang, Z.; Ma, L.; Zhang, Y.; Niu, Z. Implementation of USEPA RfD and SFO for improved risk assessment of organophosphate esters (organophosphate flame retardants and plasticizers). *Environ. Int.* **2018**, *114*, 21–26. [CrossRef]
44. Yang, J.; Luo, Y.; Chen, M.; Lu, H.; Zhang, H.; Liu, Y.; Guo, C.; Xu, J. Occurrence, spatial distribution, and potential risks of organic micropollutants in urban surface waters from qinghai, northwest China. *Chemosphere* **2023**, *318*, 137819. [CrossRef] [PubMed]
45. Tang, A.; Zhang, X.; Li, R.; Tu, W.; Guo, H.; Zhang, Y.; Li, Z.; Liu, Y.; Mai, B. Spatiotemporal distribution, partitioning behavior and flux of per- and polyfluoroalkyl substances in surface water and sediment from Poyang Lake, China. *Chemosphere* **2022**, *295*, 133855. [CrossRef] [PubMed]
46. Aravindan, V.; Gnanaraj, J.; Madhavi, S.; Liu, H.K. Lithium-ion conducting electrolyte salts for lithium batteries. *Chem.-A Eur. J.* **2011**, *17*, 14326–14346. [CrossRef] [PubMed]
47. Ober, C.K.; Käfer, F.; Deng, J. Review of essential use of fluorochemicals in lithographic patterning and semiconductor processing. *J. Micro/Nanopatterning Mater. Metrol.* **2022**, *21*, 010901. [CrossRef]
48. Zhi, Y.; Lu, X.; Munoz, G.; Yeung, L.W.Y.; De Silva, A.O.; Hao, S.; He, H.; Jia, Y.; Higgins, C.P.; Zhang, C. Environmental Occurrence and Biotic Concentrations of Ultrashort-Chain Perfluoroalkyl Acids: Overlooked Global Organofluorine Contaminants. *Environ. Sci. Technol.* **2024**, *58*, 21393–21410. [CrossRef] [PubMed]
49. Wang, H.; Dai, J.; Wei, H.; Lu, Q. Understanding technological innovation and evolution of energy storage in China: Spatial differentiation of innovations in lithium-ion battery industry. *J. Energy Storage* **2023**, *66*, 107307. [CrossRef]
50. Neuwald, I.J.; Hübner, D.; Wiegand, H.L.; Valkov, V.; Borchers, U.; Nödler, K.; Scheurer, M.; Hale, S.E.; Arp, H.P.H.; Zahn, D. Ultra-short-chain PFASs in the sources of German drinking water: Prevalent, overlooked, difficult to remove, and unregulated. *Environ. Sci. Technol.* **2022**, *56*, 6380–6390. [CrossRef] [PubMed]
51. Wee, S.Y.; Aris, A.Z. Environmental impacts, exposure pathways, and health effects of PFOA and PFOS. *Ecotoxicol. Environ. Saf.* **2023**, *267*, 115663. [CrossRef] [PubMed]
52. van der Veen, I.; de Boer, J. Phosphorus flame retardants: Properties, production, environmental occurrence, toxicity and analysis. *Chemosphere* **2012**, *88*, 1119–1153. [CrossRef] [PubMed]
53. Cristale, J.; Dantas, R.F.; De Luca, A.; Sans, C.; Esplugas, S.; Lacorte, S. Role of oxygen and DOM in sunlight induced photodegradation of organophosphorous flame retardants in river water. *J. Hazard. Mater.* **2017**, *323*, 242–249. [CrossRef] [PubMed]
54. Cai, C.; Zhang, P.; Deng, J.; Zhou, H.; Cheng, J. Ultrasensitive determination of highly polar trimethyl phosphate in environmental water by molecularly imprinted polymeric fiber headspace solid-phase microextraction. *J. Sep. Sci.* **2018**, *41*, 1104–1111. [CrossRef]

55. Wei, G.-L.; Li, D.-Q.; Zhuo, M.-N.; Liao, Y.-S.; Xie, Z.-Y.; Guo, T.-L.; Li, J.-J.; Zhang, S.-Y.; Liang, Z.-Q. Organophosphorus flame retardants and plasticizers: Sources, occurrence, toxicity and human exposure. *Environ. Pollut.* **2015**, *196*, 29–46. [CrossRef] [PubMed]
56. Liu, Y.; Chen, L.; Li, H.; Song, Y.; Yang, Z.; Cui, Y. Occurrence of organophosphorus flame retardants in Xiangjiang River: Spatiotemporal variations, potential affecting factors, and source apportionment. *Chemosphere* **2024**, *355*, 141822. [CrossRef]
57. Cristale, J.; Oliveira Santos, I.; Umbuzeiro, G.d.A.; Fagnani, E. Occurrence and risk assessment of organophosphate esters in urban rivers from Piracicaba watershed (Brazil). *Environ. Sci. Pollut. Res.* **2021**, *28*, 59244–59255. [CrossRef] [PubMed]
58. Xu, L.; Zhang, B.; Hu, Q.; Liu, Y.; Shang, T.; Zeng, X.; Yu, Z. Occurrence and spatio-seasonal distribution of organophosphate tri-and di-esters in surface water from Dongting Lake and their potential biological risk. *Environ. Pollut.* **2021**, *282*, 117031. [CrossRef]
59. Li, W.; Wang, Y.; Kannan, K. Occurrence, distribution and human exposure to 20 organophosphate esters in air, soil, pine needles, river water, and dust samples collected around an airport in New York state, United States. *Environ. Int.* **2019**, *131*, 105054. [CrossRef]
60. Li, W.; Yuan, Y.; Wang, S.; Liu, X. Occurrence, spatiotemporal variation, and ecological risks of organophosphate esters in the water and sediment of the middle and lower streams of the Yellow River and its important tributaries. *J. Hazard. Mater.* **2023**, *443*, 130153. [CrossRef]
61. Sun, H.; Mi, W.; Li, X.; Wang, S.; Yan, J.; Zhang, G. Organophosphate ester in surface water of the Pearl River and South China Sea, China: Spatial variations and ecological risks. *Chemosphere* **2024**, *361*, 142559. [CrossRef] [PubMed]
62. Pantelaki, I.; Voutsa, D. Organophosphate esters in inland and coastal waters in northern Greece. *Sci. Total Environ.* **2021**, *800*, 149544. [CrossRef] [PubMed]
63. Yan, Z.; Feng, C.; Xu, Y.; Wang, J.; Huang, N.; Jin, X.; Wu, F.; Bai, Y. Water temperature governs organophosphate ester dynamics in the aquatic food chain of Poyang Lake. *Environ. Sci. Ecotechnology* **2024**, *21*, 100401. [CrossRef] [PubMed]
64. Campo, J.; Masiá, A.; Picó, Y.; Farré, M.; Barceló, D. Distribution and fate of perfluoroalkyl substances in Mediterranean Spanish sewage treatment plants. *Sci. Total Environ.* **2014**, *472*, 912–922. [CrossRef] [PubMed]
65. Arvaniti, O.S.; Ventouri, E.I.; Stasinakis, A.S.; Thomaidis, N.S. Occurrence of different classes of perfluorinated compounds in Greek wastewater treatment plants and determination of their solid–water distribution coefficients. *J. Hazard. Mater.* **2012**, *239–240*, 24–31. [CrossRef] [PubMed]
66. Shi, B.; Wang, T.; Yang, H.; Zhou, Y.; Bi, R.; Yang, L.; Yoon, S.J.; Kim, T.; Khim, J.S. Perfluoroalkyl acids in rapidly developing coastal areas of China and South Korea: Spatiotemporal variation and source apportionment. *Sci. Total Environ.* **2021**, *761*, 143297. [CrossRef] [PubMed]
67. Gustavsson, J.; Wiberg, K.; Ribeli, E.; Nguyen, M.A.; Josefsson, S.; Ahrens, L. Screening of organic flame retardants in Swedish river water. *Sci. Total Environ.* **2018**, *625*, 1046–1055. [CrossRef]
68. Wu, J.; Junaid, M.; Wang, Z.; Sun, W.; Xu, N. Spatiotemporal distribution, sources and ecological risks of perfluorinated compounds (PFCs) in the Guanlan River from the rapidly urbanizing areas of Shenzhen, China. *Chemosphere* **2020**, *245*, 125637. [CrossRef]
69. Prevedouros, K.; Cousins, I.T.; Buck, R.C.; Korzeniowski, S.H. Sources, Fate and Transport of Perfluorocarboxylates. *Environ. Sci. Technol.* **2006**, *40*, 32–44. [CrossRef]
70. Liu, Z.; Lu, Y.; Wang, T.; Wang, P.; Li, Q.; Johnson, A.C.; Sarvajayakesavalu, S.; Sweetman, A.J. Risk assessment and source identification of perfluoroalkyl acids in surface and ground water: Spatial distribution around a mega-fluorochemical industrial park, China. *Environ. Int.* **2016**, *91*, 69–77. [CrossRef] [PubMed]
71. Young, C.J.; Furdui, V.I.; Franklin, J.; Koerner, R.M.; Muir, D.C.; Mabury, S.A. Perfluorinated acids in arctic snow: New evidence for atmospheric formation. *Environ. Sci. Technol.* **2007**, *41*, 3455–3461. [CrossRef]
72. Bach, C.C.; Vested, A.; Jørgensen, K.T.; Bonde, J.P.E.; Henriksen, T.B.; Toft, G. Perfluoroalkyl and polyfluoroalkyl substances and measures of human fertility: A systematic review. *Crit. Rev. Toxicol.* **2016**, *46*, 735–755. [CrossRef] [PubMed]
73. Kotthoff, M.; Müller, J.; Jürling, H.; Schlummer, M.; Fiedler, D. Perfluoroalkyl and polyfluoroalkyl substances in consumer products. *Environ. Sci. Pollut. Res.* **2015**, *22*, 14546–14559. [CrossRef] [PubMed]
74. Sun, H.; Li, F.; Zhang, T.; Zhang, X.; He, N.; Song, Q.; Zhao, L.; Sun, L.; Sun, T. Perfluorinated compounds in surface waters and WWTPs in Shenyang, China: Mass flows and source analysis. *Water Res.* **2011**, *45*, 4483–4490. [CrossRef] [PubMed]
75. Huang, M.; Zeng, L.; Wang, C.; Zhou, X.; Peng, Y.; Shi, C.; Wang, S.; Li, Y.; Barceló, D.; Li, H. A comprehensive and quantitative comparison of organophosphate esters: Characteristics, applications, environmental occurrence, toxicity, and health risks. *Crit. Rev. Environ. Sci. Technol.* **2024**, *55*, 310–333. [CrossRef]
76. Zhou, L.; Hiltscher, M.; Gruber, D.; Püttmann, W. Organophosphate flame retardants (OPFRs) in indoor and outdoor air in the Rhine/Main area, Germany: Comparison of concentrations and distribution profiles in different microenvironments. *Environ. Sci. Pollut. Res.* **2017**, *24*, 10992–11005. [CrossRef]

77. Pantelaki, I.; Voutsa, D. Organophosphate flame retardants (OPFRs): A review on analytical methods and occurrence in wastewater and aquatic environment. *Sci. Total Environ.* **2019**, *649*, 247–263. [CrossRef]
78. Camacho, C.G.; Antonison, A.; Oldnettle, A.; Costa, K.A.; Timshina, A.S.; Ditz, H.; Thompson, J.T.; Holden, M.M.; Sobczak, W.J.; Arnold, J.; et al. Statewide Surveillance and Mapping of PFAS in Florida Surface Water. *ACS ES&T Water* **2024**, *4*, 4343–4355. [CrossRef]
79. Yeung, L.W.Y.; Yamashita, N.; Taniyasu, S.; Lam, P.K.S.; Sinha, R.K.; Borole, D.V.; Kannan, K. A survey of perfluorinated compounds in surface water and biota including dolphins from the Ganges River and in other waterbodies in India. *Chemosphere* **2009**, *76*, 55–62. [CrossRef]
80. Ma, Y.; Luo, Y.; Zhu, J.; Zhang, J.; Gao, G.; Mi, W.; Xie, Z.; Lohmann, R. Seasonal variation and deposition of atmospheric organophosphate esters in the coastal region of Shanghai, China. *Environ. Pollut.* **2022**, *300*, 118930. [CrossRef]
81. Hoke, R.A.; Ferrell, B.D.; Ryan, T.; Sloman, T.L.; Green, J.W.; Nabb, D.L.; Mingoia, R.; Buck, R.C.; Korzeniowski, S.H. Aquatic hazard, bioaccumulation and screening risk assessment for 6: 2 fluorotelomer sulfonate. *Chemosphere* **2015**, *128*, 258–265. [CrossRef]
82. Hoke, R.A.; Bouchelle, L.D.; Ferrell, B.D.; Buck, R.C. Comparative acute freshwater hazard assessment and preliminary PNEC development for eight fluorinated acids. *Chemosphere* **2012**, *87*, 725–733. [CrossRef]
83. Ulhaq, M.; Carlsson, G.; Örn, S.; Norrgren, L. Comparison of developmental toxicity of seven perfluoroalkyl acids to zebrafish embryos. *Environ. Toxicol. Pharmacol.* **2013**, *36*, 423–426. [CrossRef] [PubMed]
84. Valsecchi, S.; Conti, D.; Crebelli, R.; Polesello, S.; Rusconi, M.; Mazzoni, M.; Preziosi, E.; Carere, M.; Lucentini, L.; Ferretti, E. Deriving environmental quality standards for perfluorooctanoic acid (PFOA) and related short chain perfluorinated alkyl acids. *J. Hazard. Mater.* **2017**, *323*, 84–98. [CrossRef]
85. Hu, J.; Lyu, Y.; Chen, H.; Cai, L.; Li, J.; Cao, X.; Sun, W. Integration of target, suspect, and nontarget screening with risk modeling for per-and polyfluoroalkyl substances prioritization in surface waters. *Water Res.* **2023**, *233*, 119735. [CrossRef]
86. Leng, Y.; Xiao, H.; Li, Z.; Liu, Y.; Huang, K.; Wang, J. Occurrence and ecotoxicological risk assessment of perfluoroalkyl substances in water of lakes along the middle reach of Yangtze River, China. *Sci. Total Environ.* **2021**, *788*, 147765. [CrossRef]
87. Li, J.; Ai, Y.; Hu, J.; Xu, N.; Song, R.; Zhu, Y.; Sun, W.; Ni, J. Polyfluoroalkyl substances in Danjiangkou Reservoir, China: Occurrence, composition, and source appointment. *Sci. Total Environ.* **2020**, *725*, 138352. [CrossRef]
88. *Test No. 201*; Freshwater Alga and Cyanobacteria, Growth Inhibition Test. OECD: Paris, France, 2011. [CrossRef]
89. *Test No. 202*; *Daphnia* sp. Acute Immobilisation Test. OECD: Paris, France, 2004. [CrossRef]
90. *Test No. 203*; Fish, Acute Toxicity Test. OECD: Paris, France, 2019. [CrossRef]
91. Gebreab, K.Y.; Benetti, D.; Grosell, M.; Stieglitz, J.D.; Berry, J.P. Toxicity of perfluoroalkyl substances (PFAS) toward embryonic stages of mahi-mahi (*Coryphaena hippurus*). *Ecotoxicology* **2022**, *31*, 1057–1067. [CrossRef] [PubMed]
92. Berends, A.G.; Boutonnet, J.C.; Rooij, C.G.D.; Thompson, R.S. Toxicity of trifluoroacetate to aquatic organisms. *Environ. Toxicol. Chem. Int. J.* **1999**, *18*, 1053–1059. [CrossRef]
93. Garavagno, M.d.l.A.; Holland, R.; Khan, M.A.H.; Orr-Ewing, A.J.; Shallcross, D.E. Trifluoroacetic Acid: Toxicity, Sources, Sinks and Future Prospects. *Sustainability* **2024**, *16*, 2382. [CrossRef]
94. Wang, G.; Shi, H.; Du, Z.; Chen, H.; Peng, J.; Gao, S. Bioaccumulation mechanism of organophosphate esters in adult zebrafish (Danio rerio). *Environ. Pollut.* **2017**, *229*, 177–187. [CrossRef]
95. Yi, Y.; Yang, Z.; Zhang, S. Ecological risk assessment of heavy metals in sediment and human health risk assessment of heavy metals in fishes in the middle and lower reaches of the Yangtze River basin. *Environ. Pollut.* **2011**, *159*, 2575–2585. [CrossRef] [PubMed]

toxics

MDPI

Article

The Effects of Disinfection Byproduct 2,6-Dichloro-1,4-benzoquinone on the Cyanobacterium *Microcystis aeruginosa*: From the Perspectives of Biochemistry and Non-Targeted Metabolomics

Tianqi Zhang [1,†], Zhaoyang Wang [1,†], Liang Wu [2], Chaonan Liu [1], Liang Meng [3,4], Fuxiang Tian [1], Meifang Hou [1], Haizhuan Lin [5,*] and Jing Ye [1,6,*]

1 School of Chemical and Environmental Engineering, Shanghai Institute of Technology, Shanghai 201418, China; 236061257@mail.sit.edu.cn (T.Z.); 15850606769@163.com (Z.W.); lcn@sit.edu.cn (C.L.); 13916841275@163.com (F.T.); mfhou@sit.edu.cn (M.H.)
2 Department of Chemical and Environmental Engineering, University of California, Riverside, CA 92521, USA; chriswu1009@hotmail.com
3 School of Environmental and Geographical Sciences, Shanghai Normal University, Shanghai 200234, China; mengliang@shnu.edu.cn
4 Yangtze River Delta Urban Wetland Ecosystem National Field Scientific Observation and Research Station, Shanghai 201722, China
5 College of Architecture and Energy Engineering, Wenzhou University of Technology, Wenzhou 325000, China
6 State Key Laboratory of Environmental Criteria and Risk Assessment, Chinese Research Academy of Environmental Sciences, Beijing 100012, China
* Correspondence: linhaizhuan@wzut.edu.cn (H.L.); yejinganna@sit.edu.cn (J.Y.)
† These authors contributed equally to this work and should be considered co-first authors.

Abstract: 2,6-Dichloro-1,4-benzoquinone (2,6-DCBQ) is an emerging chlorinated disinfection byproduct (DBP) in bodies of water. However, this compound poses an unknown toxic effect on cyanobacteria. In this study, the toxicological mechanisms of 2,6-DCBQ in *Microcystis aeruginosa* (*M. aeruginosa*) were investigated through physiological and nontargeted metabolomic assessments. The results show that 2,6-DCBQ inhibited the growth of *M. aeruginosa*, reduced its photosynthetic pigment and protein contents, increased the levels of reactive oxygen species, damaged the antioxidant defense system, and aggravated the cytomembrane. Meanwhile, 2,6-DCBQ stimulated the production and release of microcystin-LR (MC-LR) and altered the transcripts of genes associated with its synthesis (*mcyA*, *mcyD*) and transport (*mcyH*). In addition, nontargeted metabolomics of *M. aeruginosa* cells exposed to 0.1 mg/L 2,6-DCBQ identified 208 differential metabolites belonging to 10 metabolic pathways and revealed the considerable interference caused by 2,6-DCBQ among ABC transporters, the two-component system, and folate biosynthesis. This study deepens the understanding of the physiological and nontargeted metabolomic responses of *M. aeruginosa* exposed to 2,6-DCBQ, offers insights into the toxic effect of 2,6-DCBQ on *M. aeruginosa*, and provides a theoretical basis for the ecological risk assessment of emerging DBPs in accordance with water quality criteria.

Keywords: 2,6-dichloro-1,4-benzoquinone; *Microcystis aeruginosa*; toxic effect; non-targeted metabolomics; water quality criteria

1. Introduction

The COVID-19 pandemic spread globally and greatly threatened human health [1]. The worldwide COVID-19 pandemic has resulted in a substantial rise in the use of disinfectants around the globe. Excessive use of disinfectants leads to increased concentrations of residual chlorine and disinfection byproducts (DBPs) in water bodies [2]. Thus far, DBPs have been extensively detected in a variety of reclaimed and surface waters [3]. Although the environmental levels of DBPs (~μg/L) are typically considerably lower than their effective toxic concentrations (~mg/L), recent studies have shown that DBPs at environmentally relevant concentrations can also pose environmental risks [4,5]. At present, over 900 DBPs have been detected [6], but only a limited number have been investigated.

Halobenzoquinones (HBQs), a new class of DBPs [7,8], have garnered research interest due to studies demonstrating their potential cytotoxicity, genotoxicity, and carcinogenicity [9,10]. Toxicological experiments have confirmed the high cytotoxicity of HBQs, with toxicity levels approximately 10–1000 times higher than those of regulated DBPs, such as trichloromethane and halogenated acetic acids [7]. 2,6-Dichloro-1,4-benzoquinone (2,6-DCBQ) is the most frequently detected type of HBQs, with the highest detection concentration of 274.5 $\pm$ 13.0 ng/L [11,12]. The toxic effect of 2,6-DCBQ is an important issue in the research on DBPs. Previous research has indicated that 2,6-DCBQ-induced toxicity is mainly due to the presence of reactive oxygen species (ROS) [13]. However, the toxicity and potential mechanisms of 2,6-DCBQ have not been explored, especially in aquatic ecosystems.

Cyanobacterial blooms have attracted worldwide attention in the past several decades [14]. Cyanobacterial blooms have caused detrimental effects on the ecological safety of water environments, including the restriction of light penetration, the depletion of dissolved oxygen, and the mortality of aquatic organisms [15–17]. In addition, eutrophic lakes have encountered new challenges due to the introduction of emerging contaminants [18,19]. *Microcystis aeruginosa* is the most common type of phytoplankton [20,21]. *M. aeruginosa*, as a primary producer in aquatic habitats, is a major contributor to global biogeochemical cycles [22]. However, as a potential toxic cyanobacterium, it produces microcystins, which exhibit significant toxicity to other aquatic organisms. Cyanobacteria that have been exposed to environmental stressors experience physiological changes, which can be used in the effective identification of the toxic effects of pollutants and the further unveiling of their underlying mechanisms. The emerging contaminants (such as DBPs) released into the aquatic environment may contribute to the outbreak of cyanobacterial blooms through their effects on cyanobacterial growth and metabolism [23,24]. Therefore, conducting studies on the impact of emerging contaminants, such as 2,6-DCBQ, on *M. aeruginosa* has profound significance.

This study investigated the physiological response of *M. aeruginosa* exposed to 2,6-DCBQ. The cell density and contents of chlorophyll (Chl)-a, carotenoids, adenosine triphosphate (ATP), proteins, and glucose were examined. Measurements were conducted for the ROS level, microcystin-LR (MC-LR) content, lipid peroxidation, and activities of antioxidant-related enzymes in *M. aeruginosa*. In addition, observations were focused on the morphology and ultrastructural characteristics of cyanobacteria cells. Finally, non-targeted metabolomics was applied in the metabolite analysis and identification of toxicity biomarkers. The results can aid in the elucidation of the toxicity mechanisms of 2,6-DCBQ on *M. aeruginosa* and provide a theoretical basis for the ecological risk assessment of emerging DBPs in accordance with water quality criteria.

2. Materials and Methods

2.1. Chemicals and M. aeruginosa Cultivation

2,6-Dichloro-1,4-benzoquinone (2,6-DCBQ) (CAS No. 697-91-6) was obtained from TCL Co., Ltd. (Shanghai, China). An MC-LR standard sample (purity $\geq$ 95%) was obtained from Puhuaren Biological Development Co., Ltd. (Beijing, China).

M. aeruginosa (FACHB-905) was acquired from the Freshwater Algae Culture collection at the Institute of Hydrobiology, Chinese Academy of Sciences (Wuhan, China). The cultivation conditions were referenced from our previous paper [25]. Based on the preliminary experiments, exposure concentrations of 2,6-DCBQ were set as follows: 0.01, 0.05, 0.1, 0.2, and 0.5 mg/L. According to the OECD Test Guideline No. 201, dimethyl sulfoxide (purity $\geq$ 99.9%) was used as a carrier to dissolve 2,6-DCBQ, which was added to the *M. aeruginosa* culture media at a volume below 0.01% (v/v).

2.2. Measurements of Growth and the Contents of Chl-a, Carotenoid, and Protein

Based on preliminary experiments, five concentrations (0.01, 0.05, 0.1, 0.2, and 0.5 mg/L) were selected for testing. Guava flow cytometry (Luminex, Austin, TX, USA) was used to acquire the standard curve correlating the cell density with the optical density. The measurements were analyzed using the following equations [25]:

$$\text{Cell density} = 13.77488 \times \text{OD680} + 0.75405\ (R^2 = 0.9934) \tag{1}$$

We determined the content of chlorophyll a and carotenoids in the *M. aeruginosa* using the N, N-dimethylformamide method. Each group of *M. aeruginosa* cells were collected on days 3, 6, and 9 for the chl-a and carotenoid content measurements. The measurements were analyzed using the following equations:

$$\text{Chl-a} = 13.7 \times \text{OD665} - 5.76 \times \text{OD649} \tag{2}$$

$$\text{Carotenoid} = (1000 \times \text{OD470} - 1.91\text{chl-a})/225 \tag{3}$$

The intracellular protein content was measured according to the instructions of the kit (A045-2-2, Jiancheng Bioengineering Institute, Nanjing, China) on days 3 and 6, which are based on the basic principles of the Bradford method.

2.3. ATP Content and Glucose Content Detection

The ATP content was determined according to phosphomolybdic acid colorimetry. The ATP content was measured according to the instructions of the kit (A095-1-1, Jiancheng Bioengineering Institute, Nanjing, China) on days 3 and 6. The luminescence intensity was measured at 636 nm. The glucose content was determined according to the glucose oxidase method. The glucose content was measured according to the instructions of the kit (F006-1-1, Jiancheng Bioengineering Institute, Nanjing, China) on days 3 and 6.

2.4. Determination of Oxidative Stress Biomarkers

After 24 and 48 h of exposure, the ROS level was measured according to the instructions of the kit (A106-1-2), which are based on the basic principles of 2′,7′-Dichlorodihydrofluorescein diacetate. After 24 and 48 h of exposure, the lipid hydroperoxide (LPO) content was measured according to the instructions of the kit (A106-1-2), which are based on the basic principles of the colorimetric method. The catalase (CAT) and superoxide dismutase (SOD) activities were determined according to the ammonium molybdate method and a xanthine oxidase assay, respectively. The CAT and SOD activities were measured according to the instructions of the kits (A007-1-1 and A001-3-2). The operation and calculation procedures were conducted in accordance with the instructions provided by the kit (Jiancheng Bioengineering Institute, Nanjing, China).

2.5. Scanning Electron Microscope (SEM) and Transmission Electron Microscope (TEM) Observations

SEM and TEM observations were conducted according to our previous studies [25]. After 72 h of exposure, 1mL of the cell sample was centrifuged at 4000× g (4 °C, 8 min), and the precipitated cells were collected. The sediment was suspended in 1.5mL PBS.

The samples were fixed in 25% glutaraldehyde solution at 4 °C in the dark for 12 h, and washed three times with 0.1 M PBS (pH = 7.0). Then, the solution was dehydrated in different concentration gradients of ethanol for 15 min each and treated with 100% ethanol for 20 min. The samples were soaked in 100% propylene oxide for 20 min, then treated with a mixture of embedding agent and acetone, and embedded in epoxy resin. The surface morphology of the cells was observed using SEM at 3.0 kV with an SU8010 instrument (HITACHI, SU8010, Hitachi, Ltd., Tokyo, Japan), while the ultrastructure of the cells was observed using TEM (HITACHI H-7650, Hitachi, Ltd., Tokyo, Japan).

2.6. Detection of Microcystin-LR in *M. aeruginosa*

The MC-LR content was measured according to our previous studies [25]. After 72 and 144 h of exposure, 40 mL of culture media was centrifuged at 4000× *g* (4 °C, 8 min). The supernatant was collected as extracellular MC-LR. The pellet was utilized as an intracellular solution for MC-LR extraction. The precipitate was dissolved in 10 mL of 50% methanol and subjected to sonication at 40% power using an ultrasonic cell disruptor in an ice-water bath (below 4 °C) for 8 min. The supernatant obtained after centrifugation at 7000× *g* (4 °C, 10 min) was used for further measurement of the intracellular MC-LR.

The samples gathered underwent solid-phase extraction (SPE) and analyzed by HPLC (Shimadzu, TC-20A, Shanghai, China) [26]. The SPE cartridge used contained C18 silica gel (TS208-001, Shanghai Titan Technology Co., Ltd., Shanghai, China), with a packing material of 500 mg/6 mL per cartridge. The extraction process was divided into four steps: activation (10 mL of pure methanol and 10 mL of distilled water), sample injection, leaching (10 mL of 20% methanol), and elution (5mL of methanol). After the SPE process, the resulting eluate was dried using nitrogen at a temperature of 40 °C until it was almost dry and resuspended in 1 mL of chromatography-grade methanol. The eluate was then stored at a temperature of −20 °C for analysis. The samples were measured by HPLC (UV detector, 238 nm). The mobile phase consisted of a mixture of water and methanol at a volumetric ratio of 40:60 (*v*:*v*). The injection volume was 10 μL, with a flow rate of 1.0 mL/min. The column temperature was maintained at 35 °C, and each sample was analyzed within 15 min. The limit of determination of MC-LR is 100 μg/L.

2.7. Extraction of Metabolites and Nontargeted Metabolomics Analysis

Extraction and analysis of the metabolites were conducted by Shanghai Luming Biotechnology Co., Ltd. (Shanghai, China). To analyze the metabolites, 2 mL of the cultural samples from the CON (control group) and the 0.1 mg/L group on day 9 were collected and centrifuged at 100× *g* for 5 min [27,28]. The pellets were gently mixed and washed with 1 mL of PBS, and the cells of *M. aeruginosa* were immediately frozen in liquid nitrogen for further analysis. Then, the samples were analyzed through ultra-performance liquid chromatography–mass spectrometry (ACQUITY UPLC-MS, Waters Corporation, Milford, MA, USA) equipped with the ACQUITY UPLC BEN C18 column (100 mm × 2.1 mm, 1.7 μm; Waters Corporation). The mobile phases consisted of 0.1% formic acid in water/acetonitrile (95/5, *v*/*v*, solvent A) and 0.1% formic acid in acetonitrile/isopropanol/water (47.5/47.5/5, *v*/*v*, solvent B). The sample injection volume was 5 μL, and the flow rate was set to 0.4 mL/min. The column temperature was maintained at 45 °C.

Metabolites were identified on the metabolome database. The online platform OE-Cloud tools (https://cloud.oebiotech.com, accessed on 14 January 2025) was used to summarize and analyze the data in bioinformatics, accessed on 5 March 2024. The roles of these metabolites and metabolic pathways were analyzed using the KEGG database.

2.8. Quantitative Real-Time PCR Analysis

Quantitative real-time PCR analysis was conducted according to our previous studies [25]. The fold change in gene expression was determined using the $2^{-\Delta\Delta Ct}$ method. Cultural samples from the CON and the 0.1 mg/L 2,6-DCBQ treatment were collected on day 3. The primers of the target genes (*mcyA, mcyD, mcyH, rbcL, ftsH, recA, bchL, pckA, pgk,* and *psaB*) are listed in Table S1.

2.9. Reproducibility of the Results and Statistical Analysis

According to the OECD Test Guideline No. 201, we conducted 3 independent experiments, each with 3 replicates. Statistical analysis was performed using GraphPad Prism 8.0.1 (GraphPad Software, San Diego, CA, USA) and SPSS 27.0 (SPSS Inc., Chicago, IL, USA) to determine the significance among the treatments. One-way analysis of variance (ANOVA) was used to determine the differences between the control and treatment groups, and $p < 0.05$ was considered statistically significant. In addition, the statistical methods for nontargeted metabolomics included principal component analysis (PCA), partial least squares discriminant analysis (OPLS-DA), metabolic pathway enrichment analysis, and heatmap clustering analysis.

3. Results

3.1. Physiological Responses of M. aeruginosa Exposed to 2,6-DCBQ

3.1.1. Inhibitions of Growth, Photosynthetic Pigments, and Protein Contents

Figure 1a shows the response of growing *M. aeruginosa* to various concentrations of 2,6-DCBQ. High concentrations of 2,6-DCBQ (0.2 and 0.5 mg/L) significantly inhibited the growth of *M. aeruginosa*. Additionally, 0.05 and 0.1 mg/L 2,6-DCBQ also exhibited significant inhibition effects on day 4 ($p < 0.05$). With increasing concentration and exposure time, concentration- and time-dependent effects were observed.

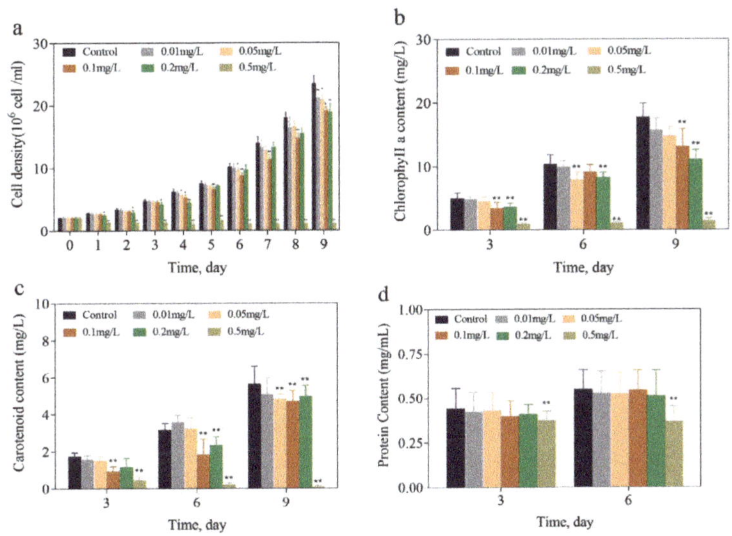

Figure 1. Cell density (**a**) of *M. aeruginosa* treated with 0.01, 0.05, 0.1, 0.2, and 0.5 mg/L of 2,6-DCBQ for 9 days. Concentrations of chlorophyll a (**b**) and carotenoids (**c**) of *M. aeruginosa* treated with mg/L 2,6-DCBQ on days 3, 6, and 9. Protein content (**d**) of *M. aeruginosa* treated with mg/L 2,6-DCBQ on days 3 and 6. Results are presented as means ± standard deviations. * Indicates $p < 0.05$; ** indicates $p < 0.01$ relative to the control by ANOVA.

Growth inhibition can be indicated by the variation in the content of photosynthetic pigments. The 2,6-DCBQ group contained considerably decreased chl-a levels after 3 and 6 days of treatment (Figure 1b). On the 9th day, 0.1, 0.2, and 0.5 mg/L 2,6-DCBQ substantially inhibited chl-a, with inhibition rates of 25.17%, 36.57%, and 92.18%, respectively. The results indicate the inhibited or disrupted photosynthetic system of *M. aeruginosa*, which in turn possibly affected its growth and proliferation. In addition, the content of carotenoids significantly decreased during exposure to 0.1 and 0.5 mg/L 2,6-DCBQ on day 3, with inhibition rates of 47.06% and 72.94%, respectively (Figure 1c). In addition, 0.5 mg/L 2,6-DCBQ resulted in substantial inhibitions in the total soluble protein content, with inhibition rates of 15.9% and 35.7% on days 3 and 6, respectively (Figure 1d).

3.1.2. ATP and Glucose Contents

Figure 2e,f display the ATP and glucose contents of *M. aeruginosa* exposed to 2,6-DCBQ on days 3 and 6, respectively. The contents of ATP decreased when *M. aeruginosa* was exposed to 0.2 mg/L 2,6-DCBQ for 3 days. In addition, the ATP content significantly increased ($p < 0.01$) when *M. aeruginosa* was exposed to 0.1 and 0.5 mg/L 2,6-DCBQ for 3 and 6 days. The glucose content considerably declined with the increase in 2,6-DCBQ concentrations.

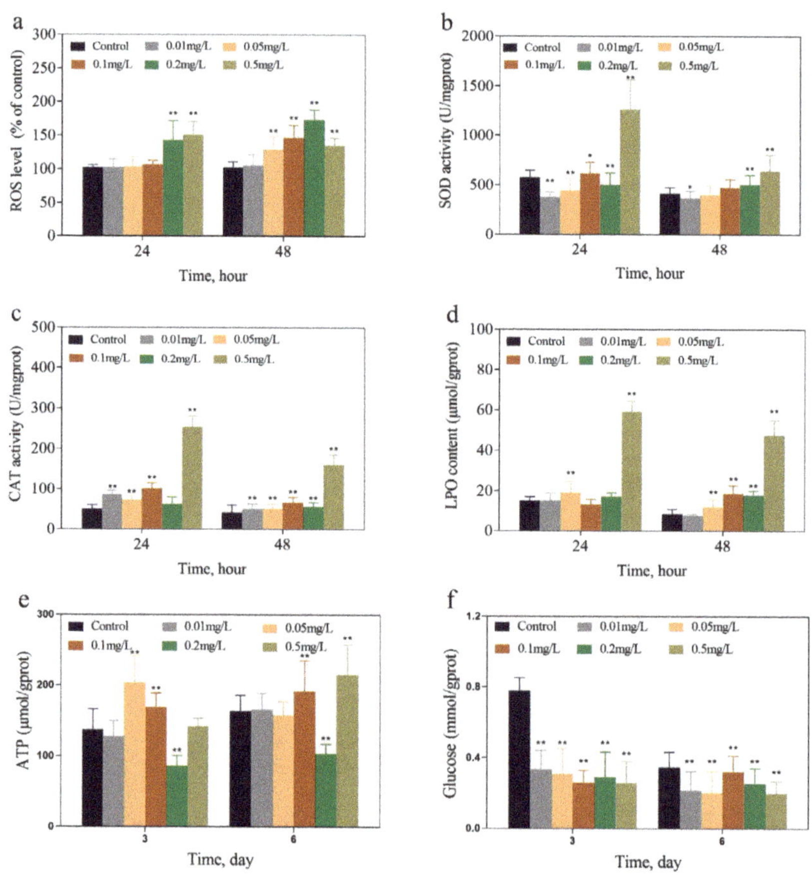

Figure 2. ROS level (**a**), SOD activity (**b**), CAT activity (**c**), and LPO content (**d**) of *M. aeruginosa* treated with 2,6-DCBQ on 24 and 48 h. ATP (**e**) and glucose (**f**) of *M. aeruginosa* treated with 2,6-DCBQ on days 3 and 6. Results are presented as means ± standard deviations. * Indicates $p < 0.05$; ** indicates $p < 0.01$ relative to the control by ANOVA.

3.1.3. ROS Level, LPO Content, and Detection of Antioxidant Activities

The antioxidant activities were evaluated to investigate the oxidative stress induced by 2,6-DCBQ in *M. aeruginosa*. The intracellular ROS level of *M. aeruginosa* increased significantly during exposure to 0.1 and 0.2 mg/L 2,6-DCBQ (Figure 2a).

Figure 2b,c show the activities of superoxide dismutase (SOD) and catalase (CAT) in *M. aeruginosa* exposed to 2,6-DCBQ. The activities of SOD and CAT presented concentration-related effects. The 0.5 mg/L 2,6-DCBQ increased the SOD activity on days 1 and 2. During *M. aeruginosa* exposure to 0.01, 0.05, 0.1, 0.2, and 0.5 mg/L 2,6-DCBQ, CAT activities were substantially increased by 21.9%, 22.1%, 58.5%, 34.1%, and 290% of the control on day 2 (Figure 2c), respectively. Similarly, the LPO content revealed a significant increase in *M. aeruginosa* when exposed to 0.5 mg/L of 2,6-DCBQ (Figure 2d).

3.2. Cell Morphology and Ultrastructure Characteristics

Figure 3a(1) shows that the surface structure of healthy *M. aeruginosa* cells was smooth and plump, without obvious damage or rupture. The cell membrane (CM) and cell wall (CW) were intact, and the cells exhibited round or oval shapes. However, after the treatment with 0.1 mg/L 2,6-DCBQ, the cells exhibited a shriveled surface morphology, deviating from their normal spherical shape, with a wrinkled and rough surface. The SEM images reveal the presence of abundant extracellular mucus surrounding the cells (Figure 3a(2)). The TEM images show that the CM and CW were rough and fragmented. The thylakoid of *M. aeruginosa* was dispersed (Figure 3b(2),b(3)). These results indicate that 2,6-DCBQ exerted acute toxicity effects on *M. aeruginosa* by inhibiting cell growth and photosynthesis. The disruption of CM further led to the release of intracellular substances (such as MCs) into the aquatic environment, accelerating the death of cells and polluting the water environment.

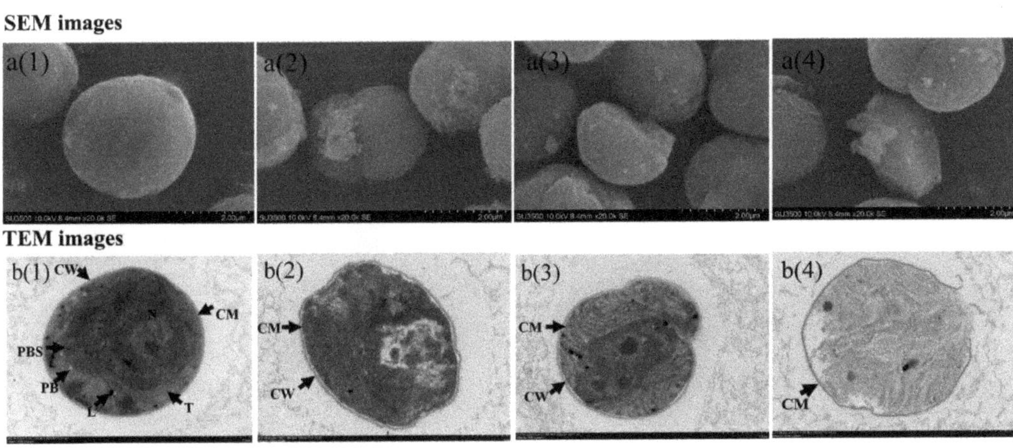

Figure 3. The effects of the CON (**a(1)**,**b(1)**) and 0.1 mg/L of 2,6-DCBQ on the ultrastructure (TEM, (**b(2)**–**b(4)**)) and surface morphology (SEM, (**a(2)**–**a(4)**)) of *M. aeruginosa* cells after 3 days of exposure. The components pictured include the nucleoid (N), phycobilisome (PBS), polyphosphate bodies (PB), thylakoids (T), leucoplast (L), cell wall (CW), and cell membrane (CM).

3.3. MC-LR Content

Figure 4 indicates that 0.1 and 0.2 mg/L 2,6-DCBQ stimulated the synthesis of MC-LR ($p < 0.01$). On days 3 and 6, the content of intracellular MC-LR increased when *M. aeruginosa* was exposed to 0.1 and 0.2 mg/L 2,6-DCBQ. Similarly, the extracellular MC-LR contents were increased when exposed to 0.1 and 0.2 mg/L 2,6-DCBQ. The results indicate that 2,6-DCBQ stimulated the production and release of MC-LR in *M. aeruginosa*.

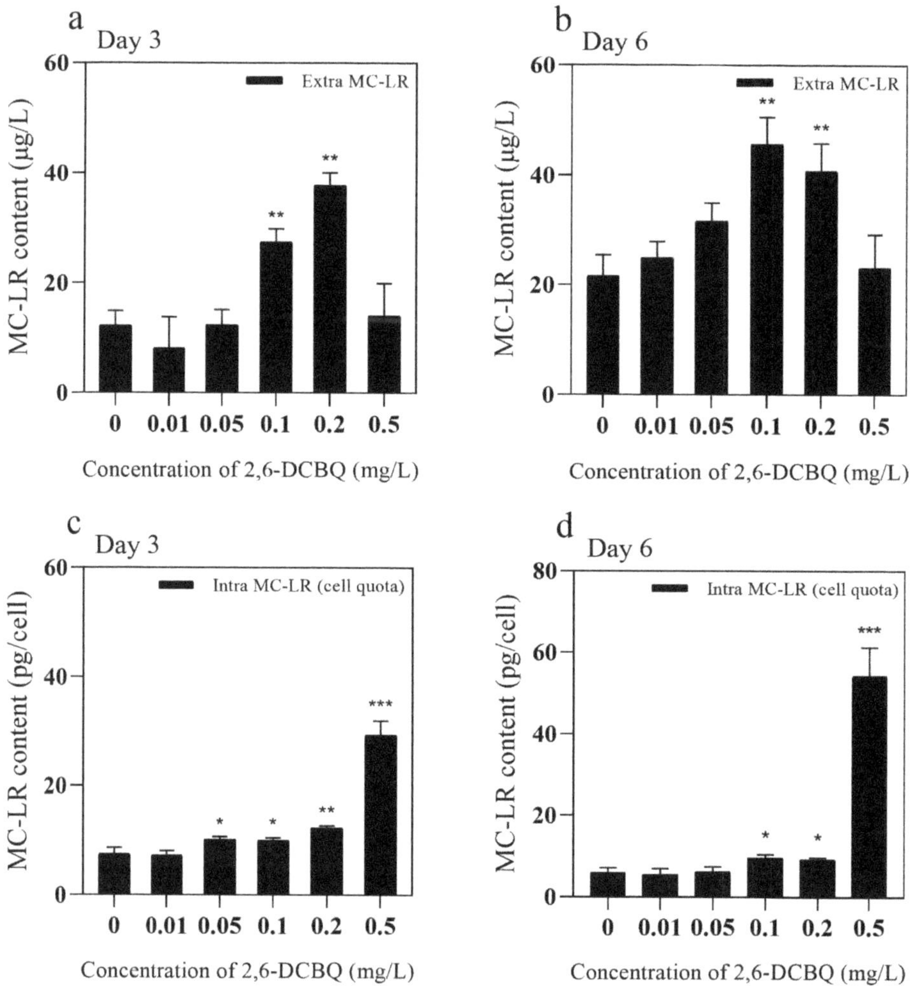

Figure 4. Effects of 2,6-DCBQ on the contents of extracellular (**a**,**b**) and intracellular (cell quota) (**c**,**d**) microcystin-LR in *M. aeruginosa*. Results are presented as means ± standard deviations. * Indicates $p < 0.05$; ** indicates $p < 0.01$; *** indicates $p < 0.001$ relative to the control by ANOVA.

3.4. Nontarget Metabolomics Analysis

3.4.1. Metabolite Detection and Analysis

We analyzed the changes in the metabolites in *M. aeruginosa* after exposure to 2,6-DCBQ. Figure 5a,b shows the score plots of principal component analysis (PCA) and orthogonal partial least squares discriminant analysis (OPLS-DA), including those of the quality control samples. The scoring plots, which are based on positive ion patterns, demonstrate the overall differences between the sample groups and the variability within each group. As shown in Figure 5a, PCA revealed distinct distributions within each group and effectively distinguished the two sample groups. The consistency of the results from the PLS-DA further supports these findings (Figure 5b).

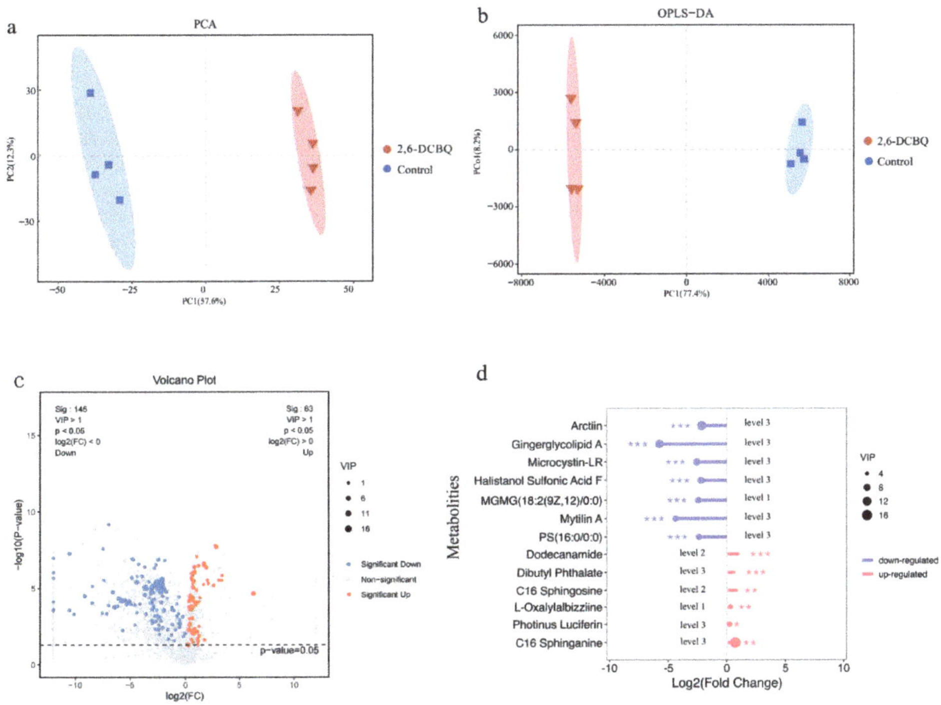

Figure 5. Two-dimensional scores plot of principal component analysis (PCA) (**a**). Two-dimensional scores plot of orthogonal partial least squares discriminant analysis (OPLS-DA) (**b**). Volcano plot of different metabolites (**c**). Lollipop map of main differential metabolites (**d**). The level indicates the accuracy of the metabolite identification, with smaller level numbers representing higher accuracy. * Indicates $p < 0.05$; ** indicates $p < 0.01$; *** indicates $p < 0.001$ relative to the control by *t* test.

3.4.2. Identification of Differential Metabolites (DMs) and Metabolic Pathway Analysis

The volcano diagram in Figure 5c illustrates the number of significantly changed metabolites during the 0.1 mg/L 2,6-DCBQ treatment of *M. aeruginosa*. In addition, the volcano plot visually display the overall distribution of the DMs. A total of 63 upregulated (red dots) and 145 downregulated (blue dots) metabolites were identified during this period (Figure 5c). Figure 5d indicates the main differential metabolites; the box plots of relative abundance are shown in Figure S1. The upregulated DMs mainly comprised organic oxygen compounds and organic nitrogen compounds (Figure 6). The downregulated DMs primarily included lipids and lipid-like molecules, organic acids, and nucleotides (Figure 6). The key enriched pathways included ABC transporters, the two-component system, and folate biosynthesis (Figure 7a). Exposure to 2,6-DCBQ can affect the KEGG pathway of *M. aeruginosa* through environmental information processing, metabolism, genetic information processing, and cellular processes (Figure 7b). ABC transporters and the two-component system are involved in environmental information processing, while folate biosynthesis is a part of metabolism (Table S2). All differential metabolites showed both upregulation and downregulation. Quorum sensing is a part of cellular processes (Table S2), with all differential metabolites being upregulated. The differential metabolites of chloroalkane and chloroalkene degradation were all upregulated, while the differential metabolites of other metabolic pathways were all downregulated.

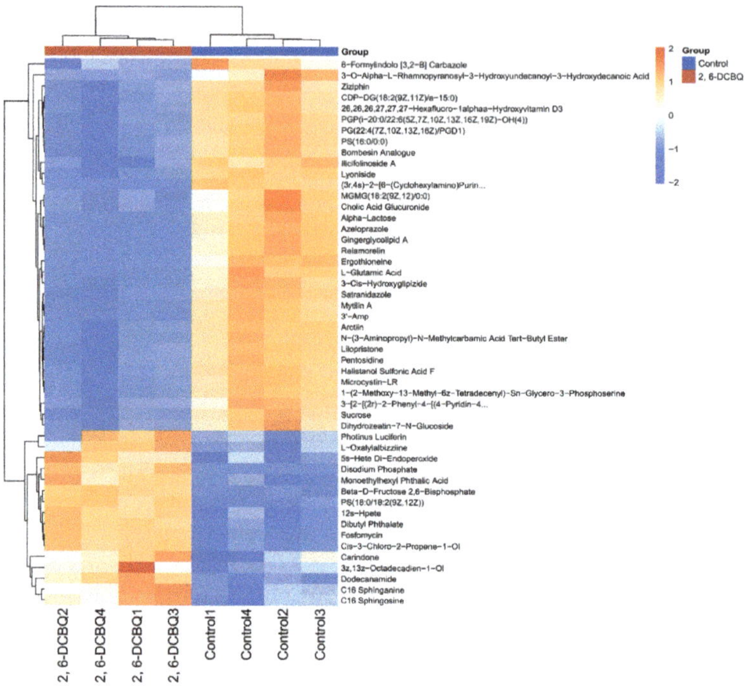

Figure 6. Cluster analysis of top 50 differential metabolites.

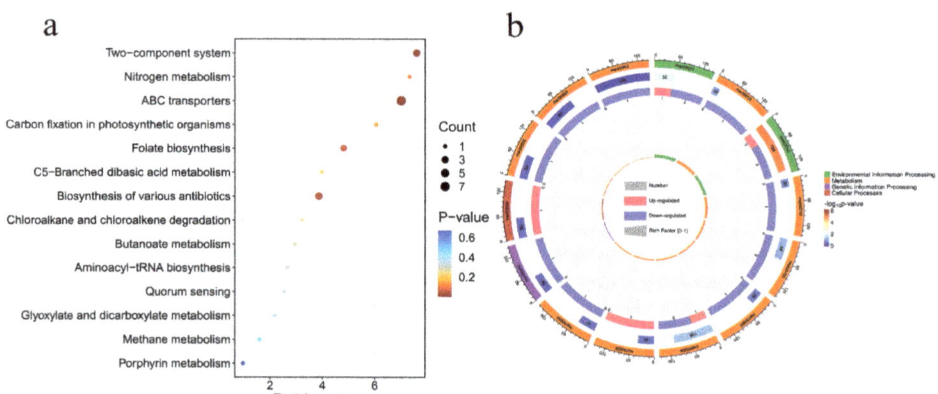

Figure 7. KEGG pathways with significant enrichment of metabolites (**a**). Circos diagram of KEGG pathways (**b**).

3.5. Gene Expressions

The expressions of genes involved in the energy metabolism, MCs, and photosynthesis were examined to explore the possible toxic mechanisms of 2,6-DCBQ. The results show that 2,6-DCBQ induced alterations in the gene expressions of *M. aeruginosa* (Figure 8). After the exposure of *M. aeruginosa* to 0.1 mg/L 2,6-DCBQ, the expressions of *mcyA*, *mcyD*, *mcyH*, *ftsH*, *recA*, *bchL*, *pckA*, and *pkg* revealed significant downregulation ($p < 0.05$). Meanwhile, the expressions of *rbcL* and *psaB* did not change significantly.

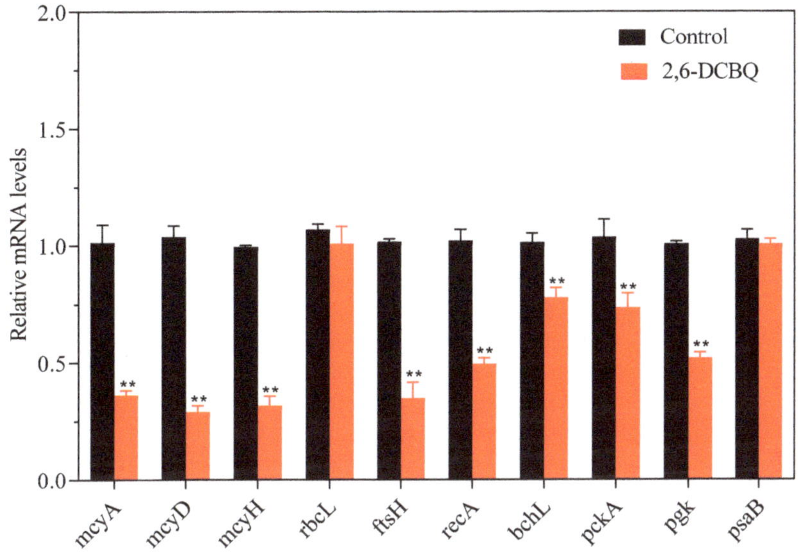

Figure 8. Gene expression levels in treatments with 0.1 mg/L 2,6-DCBQ after 3 days of exposure. Results are presented as means ± standard deviations. ** Indicates $p < 0.01$ relative to the control by ANOVA.

4. Discussion

4.1. Physiological Responses of M. aeruginosa *Exposed to 2,6-DCBQ*

4.1.1. Growth and Photosynthetic Pigments

In this study, the results on cell density after exposure to 2,6-DCBQ indicate that 2,6-DCBQ exhibits acute toxicity (Figure 1a). *recA* plays a critical role in DNA repair [29,30]. The present study reveals that the significantly decreased expression of the *recA* gene suggests DNA damage may also indicate inhibited growth of *M. aeruginosa* due to exposure to 2,6-DCBQ (Figure 8).

The influence of 2,6-DCBQ on the chl-a and carotenoid contents was similar to those on cell growth. Furthermore, the decrease in the chl-a content could impair the antioxidant capability of *M. aeruginosa*, resulting in an energy deficit in the cells and reducing their resistance to ROS-induced damage [19]. Carotenoids can react with lipid peroxides and protect the photosynthetic system of *M. aeruginosa* [31,32]. Tiwari et al. [33] also noted that the heavy metal chromium can decrease the contents of chlorophyll and carotenoids in paddy-field cyanobacteria.

The decreases in chl-a (Figure 1b) suggest the impairment or disturbance of the photosynthetic system of *M. aeruginosa*, which hindered cell growth. Li et al. [34] confirmed the same phenomenon when they exposed *Chlorella pyrenoidosa* to perfluorooctanoic acid. In summary, 2,6-DCBQ restricted the expressions of photosynthesis-related genes in *M. aeruginosa*, which resulted in its decreased photosynthetic efficiency and inhibited growth. However, the result of the *psaB* gene suggests that photosystem I of *M. aeruginosa* was not severely damaged by 0.1 mg/L 2,6-DCBQ.

4.1.2. ATP, Glucose, and Protein Contents

ATP is the direct energy source in organisms, driving various biochemical reactions and cellular activities. With prolonged exposure, the ATP content in the high-concentration treatment groups (0.1–0.5 mg/L 2,6-DCBQ) significantly increased, indicating that ATP is continuously synthesized to provide energy for repairing stress damage and other functions. A previous study also revealed that the chiral fragrance carvone affected the ATP production of *M. aeruginosa* [35], which is consistent with our experimental results.

The glucose and total soluble protein contents were comprehensively studied, and the results reveal the time–concentration toxic effect of 2,6-DCBQ on *M. aeruginosa*. In this study, reductions in the total soluble protein and glucose production were observed, indicating that disruptions in energy metabolism during photosynthesis may be involved in the growth inhibition of *M. aeruginosa* caused by 2,6-DCBQ.

The change in the total soluble protein content is a key response of *M. aeruginosa* under external stress. The total soluble protein content of M. aeruginosa decreased continuously under 2,6-DCBQ exposure (Figure 1d). The results indicate that the normal physiological activities of the cells were disrupted [35]. The *ftsH* gene plays a vital role in the maintenance of the quality of membrane proteins during the response to heat shock and in the regulation of cell division [36]. Midepogu et al. also found that after 96 h of treatment with TiO_2 nanoparticles at a concentration of 20 mg/L, the *ftsH* gene in *Chlorella pyrenoidosa* decreased to approximately 0.3-fold of the control [37]. The downregulation of the cell division protein-related *ftsH* gene may be a direct cause of the inhibitory effect on cell growth (Figure 8).

4.1.3. Oxidative Stress and Membrane Permeability

Excessive ROS production can affect various cellular processes through the alteration of nucleic acids, the oxidization of proteins, and the triggering of lipid peroxidation [38]. In this study, *M. aeruginosa* exposed to 2,6-DCBQ showed ROS accumulation and subsequent lipid peroxidation. In addition, excessive generation of ROS can initiate oxidative damage to the cell membrane, which results in morphological changes, such as disruption of cellular structures and cell death. To investigate the mechanism of oxidative damage, the effect of 2,6-DCBQ on lipid peroxidation was analyzed.

The results on the LPO content indicate that the lipids of *M. aeruginosa* were oxidatively damaged due to exposure to 2,6-DCBQ, and antioxidant enzymes cannot completely eliminate ROS [39]. In addition, a high concentration of 2,6-DCBQ may possibly induce the rise in LPO levels, which can increase cell membrane permeability [40]. The TEM results also confirm this conclusion (Figure 3b(2)). Our previous study further revealed that chloroacetic acid caused oxidative stress on *M. aeruginosa* [25].

Excessive production of SOD is the primary mechanism underlying the protection of thylakoid membranes against organic pollution [41]. Increased SOD activity protected green algae from the toxic effects of a pesticide known as trifloxystrobin [42]. The upregulation of CAT activity is considered an adaptation to the stress experienced by *M. aeruginosa*, which is stimulated by various environmental and chemical stressors [43].

4.1.4. Production and Release of MC-LR

MCs are secondary metabolites of cyanobacteria [44]. Figure 4 indicates that 2,6-DCBQ increased the production and release of MC-LR. Exposure to emerging pollutants can upregulate the expressions of MC-related genes in *M. aeruginosa*, which will promote MC-LR secretion [45].

The *mcy* gene cluster plays a key role in the biosynthesis of MCs in various cyanobacterial genera, including *M. aeruginosa*. Genes *mcyA* and *mcyD* encode the synthetase of MCs. Gene *mcyH* encodes the transportation of MCs [46]. Numerous studies have verified the relationship between the upregulation of the relative transcript abundance of *mcy* gene and the increased contents of MC-LR during production [35,47]. In addition, the common contaminant γ-lindane upregulates the transcription of *mcyD* and *mcyH* genes and improves MC production in *M. aeruginosa* [48]. However, a previous study also reported that under external stressors, such as fluoroquinolone antibiotics, the extracellular concentration of MCs increased despite inhibited *M. aeruginosa* growth, potentially due to the regulation of the *mcy* cluster [49]. In the present study, genes of *mcyA*, *mcyD*, and *mcyH* were substantially decreased compared with the control (Figure 8). These results are possibly due to severe damage of the cell membrane (Figure 3b(4)), which may result in an increase in the extracellular MC-LR content.

4.2. Nontargeted Metabolomics Analysis

Such metabolites include substances related to antioxidation (alpha-lactose, sucrose, and dihydrozeatin-7-N-glucoside), as well as nucleosides, nucleotides and their analogues, organic acids, derivatives, and lipids, all of which were altered when *M. aeruginosa* was exposed to 2,6-DCBQ (Figure 6).

Glutamic acid can be transformed into alpha-ketoglutaric acid, which is an essential intermediate in the tricarboxylic acid (TCA) cycle, through the actions of glutamate dehydrogenase, alanine aminotransferase, or aspartate aminotransferase [32]. In photosynthetic organisms, the synthesis of chl starts from glutamate, which subsequently undergoes a series of intricate biochemical reactions that involve a minimum of 17 enzyme reaction steps [50]. The stress caused by 2,6-DCBQ resulted in downregulated contents of L-glutamic acid (Figure 6), which indicates that the TCA cycle was inhibited and amino acid metabolism was disrupted. In addition, the decline in genes encoding *pckA* (Figure 8) led to the blockage of the TCA cycle. Thus, 2,6-DCBQ can hinder the energy metabolism pathway.

Lipids are the main components of organelle membranes in cyanobacteria [51]. In this study, several fatty acids (such as the potassium salts of fatty acids) in the 2,6-DCBQ exposure group were downregulated, which indicates the decreased biosynthesis of some fatty acids. In addition, the fatty acid composition of *M. aeruginosa* changes under the stress of pollutants [28,52]. Glycerophospholipids take part in the development of biological membranes and contribute to protein recognition and cell membrane signaling [53]. In this study, numerous lipid DMs, such as dodecanamide, carindone, and gingerglycolipid, were screened based on the nontargeted metabolomics of 2,6-DCBQ-treated *M. aeruginosa*. This study hypothesized that 2,6-DCBQ would affect the lipid content, which would consequently alter the selective permeability of the biomembrane and lead to the death of *M. aeruginosa*. Consequently, alterations in the types and quantities of membrane-bound lipids result in the heightened permeability of the biomembrane system. Lipid metabolism is the core pathway for cell energy and substance metabolism; it also plays a key role in basic physiological processes, such as cell membrane formation and intracellular signal transduction. Changes in these pathways inevitably lead to disturbances across the entire metabolic network, consistent with findings in previous reports. Additionally, heatmap analysis revealed the relationship between metabolites and antioxidants (Figure 6).

The pathways significantly downregulated, including ABC transporters and the two-component system ($p < 0.05$) (Figure 7a). ABC transporters, particularly ABC transporter BmrD, can participate in the absorption, accumulation, and expulsion of endogenous toxins and exogenous substances, playing an important role in defense mechanisms [51,54]. The two-component system plays a crucial role in modulating responses to environmental changes. The results show that the related signaling pathways were inhibited. The downregulation of folate biosynthesis indicates that the folate cycle was inhibited. However, due to the high diversity of metabolism species and a limited pathway database, more DMs remain to be further studied in the future.

5. Conclusions

This work reveals the physiological responses and nontargeted metabolomics of *M. aeruginosa* exposed to various concentrations of 2,6-DCBQ. The possible mechanisms underlying the mortality of *M. aeruginosa* under 2,6-DCBQ stress can be summarized as follows: (1) 2,6-DCBQ hampered the synthesis of photosynthetic pigments and further influenced photosynthesis and energy metabolism; (2) 2,6-DCBQ increased ROS production and reduced the antioxidant capacity of *M. aeruginosa*; (3) 2,6-DCBQ affected the external morphology and cell structure of *M. aeruginosa* cells; (4) 2,6-DCBQ significantly increased the production and release of MC-LR while downregulating the expression of *mcyA*, *mcyD*, and *mcyH*, possibly due to the cell membrane damage; (5) the nontargeted metabolomics analysis identified 208 differential metabolites (including antioxidants, substances related to energy metabolism, and lipids) belonging to 10 metabolic pathways. Significant metabolic enrichment pathways include ABC transporters, two-component systems, and folate biosynthesis. This study revealed the inhibitory mechanism and key biomarkers of 2,6-DCBQ on *M. aeruginosa* through in situ bioanalytical methods with multiple endpoints and nontargeted metabolomics analyses. This study also provides a theoretical basis for future research on risk assessments and the development of water quality criteria for new pollutants such as DBPs.

Supplementary Materials: The following supporting information can be downloaded at https://www.mdpi.com/article/10.3390/toxics13010064/s1: Figure S1: Box plots of relative abundance of C16, Arctiin, Gingerglycolipid A, Microcystin-Lr, Halistanol Sulfonic Acid F, Lilopristone, 1-(2-Methoxy-13-Methyl-6z-Tetradecenyl)-Sn-Glycero-, MGMG(18:2(9Z,12)/0:0), Mytilin A, PS(16:0/0:0), Photinus Luciferin, Relamorelin, Disodium Phosphate, Beta-D-Fructose 2,6-Bisphosphate, Fosfomycin, L-Oxalylalbizziine, Cis-3-Chloro-2-Propene-1-Ol, C16 Sphingosine, Dibutyl Phthalate, and Dodecanamide; Table S1: PCR primer sequences used in the experiment; Table S2: Detailed information of Circos diagram.

Author Contributions: Conceptualization, J.Y. and T.Z.; methodology, L.M.; software, Z.W.; validation, T.Z. and Z.W.; formal analysis, Z.W., C.L. and F.T.; investigation, T.Z. and Z.W.; resources, J.Y.; data curation, C.L. and F.T.; writing—original draft preparation, T.Z. and Z.W.; writing—review and editing, L.W., M.H. and H.L.; visualization, T.Z. and Z.W.; supervision, J.Y.; project administration, J.Y.; funding acquisition J.Y., L.M. and H.L. All authors have read and agreed to the published version of the manuscript.

Funding: This research was funded by the National Natural Science Foundation of China (42277274 (Jing Ye), 21307082 (Jing Ye), and 41877425 (Liang Meng)), the project of the Science and Technology Commission of Shanghai Municipality, China (18ZR1438000 (Jing Ye)), and the Study of Sources and Influence Mechanism of Nitrogen and Phosphorus Pollution in the Nanxijiang River Based on APCS-MLR Analysis (KZX202207006 (Haizhuan Lin)).

Institutional Review Board Statement: Not applicable.

Informed Consent Statement: Not applicable.

Data Availability Statement: Data will be made available upon request.

Conflicts of Interest: The authors declare no conflicts of interest.

References

1. Coronaviridae Study Group of the International Committee on Taxonomy of Viruses. The Species Severe Acute Respiratory Syndrome-Related Coronavirus: Classifying 2019-nCoV and Naming It SARS-CoV-2. *Nat. Microbiol.* **2020**, *5*, 536–544. [CrossRef]
2. Zhang, Z.; Zhou, Y.; Han, L.; Guo, X.; Wu, Z.; Fang, J.; Hou, B.; Cai, Y.; Jiang, J.; Yang, Z. Impacts of COVID-19 Pandemic on the Aquatic Environment Associated with Disinfection Byproducts and Pharmaceuticals. *Sci. Total Environ.* **2022**, *811*, 151409. [CrossRef]

3. Wang, W.; Qian, Y.; Li, J.; Moe, B.; Huang, R.; Zhang, H.; Hrudey, S.E.; Li, X.-F. Analytical and Toxicity Characterization of Halo-Hydroxyl-Benzoquinones as Stable Halobenzoquinone Disinfection Byproducts in Treated Water. *Anal. Chem.* **2014**, *86*, 4982–4988. [CrossRef]
4. He, K.; Xue, B.; Yang, X.; Wang, S.; Li, C.; Zhang, X.; Zhao, C.; Wang, X.; Qiu, Z.; Shen, Z.; et al. Low-Concentration of Trichloromethane and Dichloroacetonitrile Promote the Plasmid-Mediated Horizontal Transfer of Antibiotic Resistance Genes. *J. Hazard. Mater.* **2022**, *425*, 128030. [CrossRef] [PubMed]
5. Wang, Y.; Liu, H.; Yang, X.; Wang, L. Aquatic Toxicity and Aquatic Ecological Risk Assessment of Wastewater-Derived Halogenated Phenolic Disinfection Byproducts. *Sci. Total Environ.* **2022**, *809*, 151089. [CrossRef] [PubMed]
6. Usman, M.; Kuckelkorn, J.; Kämpfe, A.; Zwiener, C.; Wintgens, T.A.; Linnemann, V. Identification of Disinfection By-Products (DBP) in Thermal Water Swimming Pools Applying Non-Target Screening by LC-/GC-HRMS. *J. Hazard. Mater.* **2023**, *449*, 130981. [CrossRef] [PubMed]
7. Qin, F.; Zhao, Y.-Y.; Zhao, Y.; Boyd, J.M.; Zhou, W.; Li, X.-F. A Toxic Disinfection By-Product, 2,6-Dichloro-1,4-Benzoquinone, Identified in Drinking Water. *Angew. Chem.* **2010**, *49*, 790–792. [CrossRef] [PubMed]
8. Zhao, Y.; Anichina, J.; Lu, X.; Bull, R.J.; Krasner, S.W.; Hrudey, S.E.; Li, X.-F. Occurrence and Formation of Chloro- and Bromo-Benzoquinones during Drinking Water Disinfection. *Water Res.* **2012**, *46*, 4351–4360. [CrossRef]
9. Wu, H.; Long, K.; Sha, Y.; Lu, D.; Xia, Y.; Mo, Y.; Yang, Q.; Zheng, W.; Yang, M.; Wei, X. Occurrence and Toxicity of Halobenzoquinones as Drinking Water Disinfection Byproducts. *Sci. Total Environ.* **2021**, *770*, 145277. [CrossRef] [PubMed]
10. Zuo, Y.-T.; Hu, Y.; Lu, W.-W.; Cao, J.-J.; Wang, F.; Han, X.; Lu, W.-Q.; Liu, A.-L. Toxicity of 2,6-Dichloro-1,4-Benzoquinone and Five Regulated Drinking Water Disinfection by-Products for the Caenorhabditis Elegans Nematode. *J. Hazard. Mater.* **2017**, *321*, 456–463. [CrossRef] [PubMed]
11. Hu, S.; Gong, T.; Zhu, H.; Wang, J.; Li, Z.; Chen, H.; Huang, Z.; Zhang, M.; Xian, Q. Formation and Decomposition of New Iodinated Halobenzoquinones during Chloramination in Drinking Water. *Environ. Sci. Technol.* **2020**, *54*, 5237–5248. [CrossRef] [PubMed]
12. Cuthbertson, A.A.; Bach, C.; Richardson, S.D.; Dauchy, X. A Novel Automated Method for the Quantification of Ten Halobenzoquinones in Drinking Water Using Online Solid-Phase Extraction Coupled with Liquid Chromatography Tandem Mass Spectrometry. *J. Chromatogr. A* **2020**, *1612*, 460642. [CrossRef] [PubMed]
13. Liu, T.; Chen, X.; Li, W.; Zhang, X.; Wang, G.; Wang, J.; Liang, L.; Yang, F.; Li, J.; Li, J. Oxidative Stress as a Key Event in 2,6-Dichloro-1,4-Benzoquinone-Induced Neurodevelopmental Toxicity. *Ecotoxicol. Environ. Saf.* **2023**, *263*, 115357. [CrossRef]
14. Huisman, J.; Codd, G.A.; Paerl, H.W.; Ibelings, B.W.; Verspagen, J.M.H.; Visser, P.M. Cyanobacterial Blooms. *Nat. Rev. Microbiol.* **2018**, *16*, 471–483. [CrossRef] [PubMed]
15. Hallegraeff, G.; Enevoldsen, H.; Zingone, A. Global Harmful Algal Bloom Status Reporting. *Harmful Algae* **2021**, *102*, 101992. [CrossRef]
16. Yan, T.; Li, X.-D.; Tan, Z.-J.; Yu, R.-C.; Zou, J.-Z. Toxic Effects, Mechanisms, and Ecological Impacts of Harmful Algal Blooms in China. *Harmful Algae* **2022**, *111*, 102148. [CrossRef]
17. Zhang, Y.; Whalen, J.K.; Cai, C.; Shan, K.; Zhou, H. Harmful Cyanobacteria-Diatom/Dinoflagellate Blooms and Their Cyanotoxins in Freshwaters: A Nonnegligible Chronic Health and Ecological Hazard. *Water Res.* **2023**, *233*, 119807. [CrossRef] [PubMed]
18. Guo, Z.; He, H.; Yang, G.; Liu, K.; Xi, Y.; Li, Z.; Luo, Y.; Liao, Z.; Dao, G.; Ren, X.; et al. The Environmental Risks of Antiviral Drug Arbidol in Eutrophic Lake: Interactions with *Microcystis aeruginosa*. *J. Hazard. Mater.* **2024**, *466*, 133609. [CrossRef]
19. Wang, C.; Wang, R.; Hu, L.; Xi, M.; Wang, M.; Ma, Y.; Chen, J.; Liu, C.; Song, Y.; Ding, N.; et al. Metabolites and Metabolic Pathways Associated with Allelochemical Effects of Linoleic Acid on Karenia Mikimotoi. *J. Hazard. Mater.* **2023**, *447*, 130815. [CrossRef]
20. Pan, D.; Pavagadhi, S.; Umashankar, S.; Rai, A.; Benke, P.I.; Rai, M.; Saxena, G.; Gangu, V.; Swarup, S. Resource Partitioning Strategies during Toxin Production in *Microcystis aeruginosa* Revealed by Integrative Omics Analysis. *Algal Res.* **2019**, *42*, 101582. [CrossRef]
21. Zhu, X.; Dao, G.; Tao, Y.; Zhan, X.; Hu, H. A Review on Control of Harmful Algal Blooms by Plant-Derived Allelochemicals. *J. Hazard. Mater.* **2021**, *401*, 123403. [CrossRef] [PubMed]
22. Aguilera, A.; Klemenčič, M.; Sueldo, D.J.; Rzymski, P.; Giannuzzi, L.; Martin, M.V. Cell Death in Cyanobacteria: Current Understanding and Recommendations for a Consensus on Its Nomenclature. *Front. Microbiol.* **2021**, *12*, 631654. [CrossRef]
23. Wang, H.; Jin, M.; Mao, W.; Chen, C.; Fu, L.; Li, Z.; Du, S.; Liu, H. Photosynthetic Toxicity of Non-Steroidal Anti-Inflammatory Drugs (NSAIDs) on Green Algae Scenedesmus Obliquus. *Sci. Total Environ.* **2020**, *707*, 136176. [CrossRef] [PubMed]
24. Cui, H.; Zhu, X.; Zhu, Y.; Huang, Y.; Chen, B. Ecotoxicological Effects of DBPs on Freshwater Phytoplankton Communities in Co-Culture Systems. *J. Hazard. Mater.* **2022**, *421*, 126679. [CrossRef] [PubMed]
25. Ye, J.; Ni, J.; Tian, F.; Ji, X.; Hou, M.; Li, Y.; Yang, L.; Wang, R.; Xu, W.; Meng, L. Toxicity Effects of Disinfection Byproduct Chloroacetic Acid to *Microcystis aeruginosa*: Cytotoxicity and Mechanisms. *J. Environ. Sci.* **2023**, *129*, 229–239. [CrossRef] [PubMed]

26. Robillot, C.; Vinh, J.; Puiseux-Dao, S.; Hennion, M.-C. Hepatotoxin Production Kinetics of the Cyanobacterium *Microcystis aeruginosa* PCC 7820, as Determined by HPLC-Mass Spectrometry and Protein Phosphatase Bioassay. *Environ. Sci. Technol.* **2000**, *34*, 3372–3378. [CrossRef]
27. Du, C.; Sang, W.; Xu, C.; Jiang, Z.; Wang, J.; Fang, Y.; Zhu, C.; Wizi, J.; Akram, M.A.; Ni, L.; et al. Integrated Transcriptomic and Metabolomic Analysis of *Microcystis aeruginosa* Exposed to Artemisinin Sustained-Release Microspheres. *J. Hazard. Mater.* **2023**, *443*, 130114. [CrossRef] [PubMed]
28. Zhang, X.; Zhu, Q.; Chen, C.; Xie, B.; Tang, B.; Fan, M.; Hu, Q.; Liao, Z.; Yan, X. The Growth Inhibitory Effects and Non-Targeted Metabolomic Profiling of *Microcystis aeruginosa* Treated by *Scenedesmus* sp. *Chemosphere* **2023**, *338*, 139446. [CrossRef] [PubMed]
29. Wang, J.; Liu, Q.; Feng, J.; Lv, J.-P.; Xie, S.-L. Effect of High-Doses Pyrogallol on Oxidative Damage, Transcriptional Responses and Microcystins Synthesis in *Microcystis aeruginosa* TY001 (Cyanobacteria). *Ecotoxicol. Environ. Saf.* **2016**, *134*, 273–279. [CrossRef]
30. Chen, Z.; Yang, H.; Pavletich, N.P. Mechanism of Homologous Recombination from the RecA–ssDNA/dsDNA Structures. *Nature* **2008**, *453*, 489–494. [CrossRef] [PubMed]
31. Wan, L.; Wu, Y.; Zhang, B.; Yang, W.; Ding, H.; Zhang, W. Effects of Moxifloxacin and Gatifloxacin Stress on Growth, Photosynthesis, Antioxidant Responses, and Microcystin Release in *Microcystis aeruginosa*. *J. Hazard. Mater.* **2021**, *409*, 124518. [CrossRef] [PubMed]
32. Cupellini, L.; Calvani, D.; Jacquemin, D.; Mennucci, B. Charge Transfer from the Carotenoid Can Quench Chlorophyll Excitation in Antenna Complexes of Plants. *Nat. Commun.* **2020**, *11*, 662. [CrossRef]
33. Tiwari, S.; Patel, A.; Prasad, S.M. Phytohormone Up-Regulates the Biochemical Constituent, Exopolysaccharide and Nitrogen Metabolism in Paddy-Field Cyanobacteria Exposed to Chromium Stress. *BMC Microbiol.* **2020**, *20*, 206. [CrossRef]
34. Li, Y.; Liu, X.; Zheng, X.; Yang, M.; Gao, X.; Huang, J.; Zhang, L.; Fan, Z. Toxic Effects and Mechanisms of PFOA and Its Substitute GenX on the Photosynthesis of *Chlorella Pyrenoidosa*. *Sci. Total Environ.* **2021**, *765*, 144431. [CrossRef]
35. Ye, J.; Hua, S.; Liu, S.; Tian, F.; Ji, X.; Li, Y.; Hou, M.; Xu, W.; Meng, L.; Sun, L. Enantioselective Effects of Chiral Fragrance Carvone (L- and D-Carvone) on the Physiology, Oxidative Damage, Synthesis, and Release of Microcystin-LR in *Microcystis aeruginosa*. *Sci. Total Environ.* **2022**, *853*, 158631. [CrossRef] [PubMed]
36. Langklotz, S.; Baumann, U.; Narberhaus, F. Structure and Function of the Bacterial AAA Protease FtsH. *Biochim. Biophys. Acta* **2012**, *1823*, 40–48. [CrossRef]
37. Middepogu, A.; Hou, J.; Gao, X.; Lin, D. Effect and Mechanism of TiO2 Nanoparticles on the Photosynthesis of Chlorella Pyrenoidosa. *Ecotoxicol. Environ. Saf.* **2018**, *161*, 497–506. [CrossRef] [PubMed]
38. Mittler, R.; Vanderauwera, S.; Gollery, M.; Van Breusegem, F. Reactive Oxygen Gene Network of Plants. *Trends Plant Sci.* **2004**, *9*, 490–498. [CrossRef]
39. Huang, L.; Lu, D.; Diao, J.; Zhou, Z. Enantioselective Toxic Effects and Biodegradation of Benalaxyl in Scenedesmus Obliquus. *Chemosphere* **2012**, *87*, 7–11. [CrossRef] [PubMed]
40. Xu, D.; Chen, X.; Shao, B. Oxidative Damage and Cytotoxicity of Perfluorooctane Sulfonate on Chlorella Vulgaris. *Bull. Environ. Contam. Toxicol.* **2017**, *98*, 127–132. [CrossRef] [PubMed]
41. Aderemi, A.O.; Novais, S.C.; Lemos, M.F.L.; Alves, L.M.; Hunter, C.; Pahl, O. Oxidative Stress Responses and Cellular Energy Allocation Changes in Microalgae Following Exposure to Widely Used Human Antibiotics. *Aquat. Toxicol.* **2018**, *203*, 130–139. [CrossRef] [PubMed]
42. Shen, Y.-F.; Liu, L.; Gong, Y.-X.; Zhu, B.; Liu, G.-L.; Wang, G.-X. Potential Toxic Effect of Trifloxystrobin on Cellular Microstructure, mRNA Expression and Antioxidant Enzymes in Chlorella Vulgaris. *Environ. Toxicol. Pharmacol.* **2014**, *37*, 1040–1047. [CrossRef]
43. Frías-Gómez, S.A.; Hernández Hernández, L.H.; Powell, M.S.; Álvarez-González, C.A.; Cortés-Jacinto, E.; Cigarroa-Ruiz, L.; Arellano-Carrasco, G. Effect of Dietary Protein, Lipid and Carbohydrate Ratio on Growth, Digestive and Antioxidant Enzyme Activity of Prawn *Macrobrachium Acanthurus* Postlarvae. *Aquacult. Rep.* **2023**, *30*, 101578. [CrossRef]
44. Zhang, H.; Zhao, X.; Li, Y.; Xie, P. A Meta-Analysis on the Toxicity of Microcystin-LR to Fish and Mammals. *Environ. Pollut.* **2023**, *330*, 121780. [CrossRef] [PubMed]
45. Wang, Z.; Chen, Q.; Zhang, J.; Dong, J.; Ao, Y.; Wang, M.; Wang, X. Long-Term Exposure to Antibiotic Mixtures Favors Microcystin Synthesis and Release in *Microcystis aeruginosa* with Different Morphologies. *Chemosphere* **2019**, *235*, 344–353. [CrossRef]
46. Liu, Y.; Zhang, J.; Gao, B.; Feng, S. Combined Effects of Two Antibiotic Contaminants on *Microcystis aeruginosa*. *J. Hazard. Mater.* **2014**, *279*, 148–155. [CrossRef] [PubMed]
47. Liu, Y.; Wang, F.; Chen, X.; Zhang, J.; Gao, B. Influence of Coexisting Spiramycin Contaminant on the Harm of *Microcystis aeruginosa* at Different Nitrogen Levels. *J. Hazard. Mater.* **2015**, *285*, 517–524. [CrossRef] [PubMed]
48. Ceballos-Laita, L.; Calvo-Begueria, L.; Lahoz, J.; Bes, M.-T.; Fillat, M.F.; Peleato, M.-L. γ-Lindane Increases Microcystin Synthesis in *Microcystis aeruginosa* PCC7806. *Mar. Drugs* **2015**, *13*, 5666–5680. [CrossRef] [PubMed]
49. Liu, Y.; Chen, S.; Zhang, J.; Li, X.; Gao, B. Stimulation Effects of Ciprofloxacin and Sulphamethoxazole in *Microcystis aeruginosa* and Isobaric Tag for Relative and Absolute Quantitation-Based Screening of Antibiotic Targets. *Mol. Ecol.* **2017**, *26*, 689–701. [CrossRef] [PubMed]

50. Chen, M. Chlorophyll Modifications and Their Spectral Extension in Oxygenic Photosynthesis. *Annu. Rev. Biochem.* **2014**, *83*, 317–340. [CrossRef]
51. Lyu, J.; Gao, R.; Guo, Z. Galactosyldiacylglycerols: From a Photosynthesis-Associated Apparatus to Structure-Defined in Vitro Assembling. *J. Agric. Food Chem.* **2021**, *69*, 8910–8928. [CrossRef] [PubMed]
52. Chen, Y.; Zhao, C.; Zhu, X.; Zhu, Y.; Tian, R. Multiple Inhibitory Effects of Succinic Acid on *Microcystis aeruginosa*: Morphology, Metabolomics, and Gene Expression. *Environ. Technol.* **2022**, *43*, 3121–3130. [CrossRef]
53. Sato, N.; Mori, N.; Hirashima, T.; Moriyama, T. Diverse Pathways of Phosphatidylcholine Biosynthesis in Algae as Estimated by Labeling Studies and Genomic Sequence Analysis. *Plant J.* **2016**, *87*, 281–292. [CrossRef] [PubMed]
54. Reilman, E.; Mars, R.A.T.; van Dijl, J.M.; Denham, E.L. The Multidrug ABC Transporter BmrC/BmrD of Bacillus Subtilis Is Regulated via a Ribosome-Mediated Transcriptional Attenuation Mechanism. *Nucleic Acids Res.* **2014**, *42*, 11393–11407. [CrossRef] [PubMed]

toxics

MDPI

Article

Distribution of Polycyclic Aromatic Hydrocarbons and Pesticides in Danjiangkou Reservoir and Evaluation of Ecological Risk

Ruiwen Li [1,†], Hao Pang [2,†], Yemin Guo [1], Xuan Zhou [2], Kaiyu Fu [3], Taotao Zhang [4], Jian Han [3], Lihua Yang [3,*], Bingsheng Zhou [3] and Si Zhou [5,*]

1 Ecology and Environment Monitoring and Scientific Research Center, Ecology and Environment Administration of Yangtze River Basin, Ministry of Ecology and Environment, Wuhan 430010, China; liruiwen@cjjg.mee.gov.cn (R.L.); gym601027@163.com (Y.G.)

2 School of Chemistry and Environmental Engineering, Wuhan Institute of Technology, Wuhan 430074, China; 22209010140@stu.wit.edu.cn (H.P.); hbzhoux@sina.com (X.Z.)

3 Key Laboratory of Breeding Biotechnology and Sustainable Aquaculture, Institute of Hydrobiology, Chinese Academy of Sciences, Wuhan 430072, China; fuky@ihb.ac.cn (K.F.); hanjian@ihb.ac.cn (J.H.); bszhou@ihb.ac.cn (B.Z.)

4 School of Laboratory Medicine, Hubei University of Chinese Medicine, Wuhan 430065, China; vip2564@foxmail.com

5 Guizhou Institute of Environmental Science and Designing, Guiyang 550081, China

* Correspondence: lhyang@ihb.ac.cn (L.Y.); zhousi@gzhky.com (S.Z.)

† These authors contributed equally to this work.

Abstract: The Danjiangkou Reservoir is the largest artificial freshwater lake in Asia. This study investigated the spatiotemporal distribution of pesticides and polycyclic aromatic hydrocarbons (PAHs) in the Danjiangkou Reservoir to assess the ecological and human health risks associated with these pollutants. Twenty-three sampling sites in the Danjiangkou Reservoir each collected 23 surface water samples and 23 sediment samples. These samples were analyzed using gas chromatography–mass spectrometry (GC–MS), combined with risk quotient methods and health risk assessment models. The results indicated that the total concentration of PAHs (ΣPAHs) in the surface water ranged from 64.64 to 868.23 ng/L (average 217.97 ± 184.97 ng/L), and they primarily consisted of low molecular weight PAHs, with the compounds with the highest concentrations being naphthalene (10.43–116.97 ng/L), fluorene (22.74–87.61 ng/L), and phenanthrene (26.54–162.86 ng/L). The total concentration of pesticides in the surface water varied between 2.62 and 72.89 ng/L (average 22.99 ± 18.27 ng/L). In the sediment samples, the ΣPAH concentration ranged from 0.01 to 2.93 ng/g (average 0.69 ± 0.94 ng/g), and these predominantly consisted of high molecular weight PAHs, while pesticide concentrations ranged from non-detectable (nd) to 28.46 ng/g (average 7.99 ± 8.53 ng/g), with higher concentrations of malathion (0.62–9.16 ng/g) and chlorpyrifos (10.01–21.38 ng/g). Through risk assessment, it was found that although the risks posed by PAHs and pesticides to human health are very low, the ecological risk assessment indicated that certain PAHs (such as phenanthrene) and organophosphate pesticides (such as malathion and chlorpyrifos) may pose potential threats to aquatic organisms.

Keywords: Danjiangkou Reservoir; risk assessment; polycyclic aromatic hydrocarbons; organophosphorus pesticides

1. Introduction

The Danjiangkou Reservoir, which spans the provinces of Henan and Hubei, holds a distinctive position as the largest artificial freshwater lake in Asia and serves as the confluence of the Han and Dan Rivers. It plays a crucial role in the Middle Route Project of China's South-to-North Water Diversion initiative, annually supplying the substantial amount of 95 billion cubic meters of clean water to 14 major cities in northern China [1]. The land use in the vicinity of the reservoir encompasses forest land (consisting of natural forests and orchards), agricultural land (which includes dry land and paddy fields), and wetlands, as well as industrial areas. Consequently, the reservoir faces pollution threats from various human activities within its watershed, including industrial discharge, pesticide and fertilizer applications, and livestock manure [2]. Considering its strategic importance in both ecological and socio-economic development, the occurrence of organic pollutants in the aquatic environments of the Danjiangkou Reservoir and their potential harm to aquatic ecosystems and human health have received increasing attention [3,4]. Although previous studies mainly focused on water quality parameters related to eutrophication [5], occurrences of various of organic pollutants including perfluorinated substances [6], antibiotics, and microplastics have been recently detected [1,7]. According to existing studies, Polychlorinated Biphenyls (PCBs), Polybrominated Diphenyl Ethers (PBDEs), and pharmaceuticals and personal care products (PPCPs) have been detected in the Danjiangkou Reservoir. Among these, the concentration of PPCPs is relatively low, while PBDEs pose a significant risk to fish [8–10]. Habitat surveys indicate that agricultural activities are frequent near the Danjiangkou Reservoir, leading to the unavoidable use of pesticides and fuel combustion. Although industrial production and healthcare facilities are located relatively far from the reservoir, research on pollutants related to the Danjiangkou Reservoir is still relatively scarce. Current studies are insufficient to support the transition of water environment management goals from merely maintaining water quality standards to protecting ecological and health safety. On the one hand, as China is a leading agricultural nation, the use of pesticides is an important source of water pollution [11,12]. Since some traditional organochlorines (OCPs) have been prohibited in agriculture, more products have been introduced into the market, including organophosphates (OPPs) and pyrethroids. Although OPPs and pyrethroids degrade more readily than organochlorines, they may exhibit toxicity to aquatic organisms such as fish, mollusks, and arthropods, and even pose a potential threat to human health through residues in food [12] such as organophosphates pyrethroids, with some classified as global endocrine disruptors [12]. A pesticide risk assessment of Lake Taihu as well as the Jiulong River in South China found that pesticides pose a risk to the ecosystem and to people [12–14]. On the other hand, polycyclic aromatic hydrocarbons (PAHs) are a group of persistent organic pollutants (POPs) with two or more benzene rings, which primarily originate from industrial activities such as fossil fuel, petroleum development, and oil transportation, as well as some natural resources [15–17]. Polycyclic aromatic hydrocarbons (PAHs) are widely present in ecological environments [18,19]. Studies that have assessed the risk of PAHs in the Yellow River Delta and along the coast of Bohai Bay have found that PAHs pose ecological risks and harm to aquatic organisms [20,21]. Moreover, When PAHs in drinking water reach a certain concentration, they may pose a risk to human health [22], reminding us to pay attention to these pollutants.

Therefore, given the limited research on the different types of pesticides and polycyclic aromatic hydrocarbons in the region, an investigation was conducted to fill the data gap regarding these pollutants in surface water and sediment samples from the Danjiangkou Reservoir. The main objectives of this study were (1) to analyze the temporal and spatial distribution trends of polycyclic aromatic hydrocarbons and pesticides in the aquatic environment of the Danjiangkou Reservoir, and (2) to assess the ecological and human health risks of contaminants in surface water and sediments in the study area.

2. Materials and Methods

2.1. Sample Collection and Pretreatment

A total of 23 surface water samples and 23 sediment samples were collected from the Danjiangkou Reservoir in August 2020 (Figure 1), after which the surface water samples were analyzed for 10 PAHs and 11 pesticides, and the sediment samples were analyzed for 3 PAHs and 16 pesticides. D1–D11 are sites in the Dan Reservoir, H1–H7 are in the Han Reservoir. R1 is the Han River, R2 is the Si River, R3 is the Jian River, R4 is the Guanshan River, and R5 is the Langhe River in the tributaries of the Danjiangkou Reservoir.

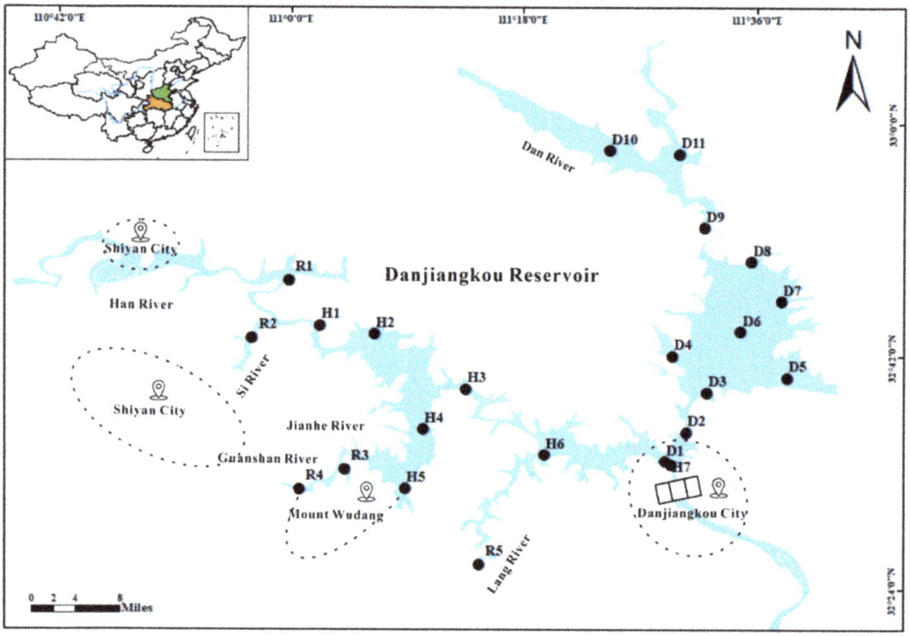

Figure 1. A map showing the sampling sites in the Danjiangkou Reservoir in Hubei Province.

Specifically, surface water samples were collected using 1-L amber glass bottles, with one liter collected per sample, and stored at −4 °C. Within 48 h, the water samples underwent filtration through 3 μm, 0.45 μm, and 0.22 μm microporous filter membranes. After filtration, 50 mL of methanol was added to the samples, which were then stored at 4 °C and the enrichment process was completed within two hours. Enrichment of the water samples was achieved through solid phase extraction (SPE), following established methods [23–25].

Sediment samples (200 to 300 g) were collected using a Petersen mud collector, cleared of weeds and debris, placed in washed polythene self-sealing bags, sealed tightly, and stored at −20 °C. The cryopreserved sediment was subsequently dried, homogenized, and sieved through a 200-mesh sieve before being stored again at −20 °C. For extraction, 10 g of the sediment sample was weighed and transferred to the Q-Cup of a fully automated pressurized fluid extraction system. The extraction was conducted at 120 °C using 25 mL of a 1:1 acetone/n-hexane mixture (comprising 20 mL of the upper layer and 5 mL of the lower layer) and 10 mL of rinsing solvent (also 1:1 acetone/n-hexane), with a dwell time of 3 min. Following this, the sample was cleaned twice with 20 mL of dichloromethane. After extraction, we carefully removed the receiver tube and placed the extract on a nitrogen blower. We used a gentle nitrogen flow rate to concentrate the extract down to 1 mL. Purification involved passing the extract through a graphitized carbon black solid-phase extraction column with an aminopropyl-bonded silica gel solid-phase extraction column in series below, both pre-drenched with 4 mL of a toluene-acetonitrile mixture. One gram of anhydrous sodium sulfate was added to the column. The sample was then loaded onto the clean-up column and drenched with 6 mL of the toluene-acetonitrile mixture. The collected drenched solution was concentrated nearly to dryness using a nitrogen blower at a moderate nitrogen flow rate. The solvent was then exchanged for dichloromethane by adding 10 μL of an internal standard solution at a concentration of 400 ppb. The final volume was adjusted to 100 μL for analysis. For further details, please refer to the published article by Black et al. (2023) [26].

2.2. Instrumental Analysis

The determination of pesticides in the extracts was conducted using gas chromatography–mass spectrometry (GC–MS) (6890N-5975, Agilent, Santa Clara, CA, USA) with a DB-5 ms (30 m) column (Agilent, Santa Clara, CA, USA). The GC conditions involved an initial hold at 40 °C for 1 min, followed by a ramp to 130 °C at 30 °C/min, another ramp to 250 °C at 5 °C/min, a further ramp to 280 °C at 10 °C/min, and a final hold at 280 °C for 8 min. The inlet temperature was set to 270 °C, with a shunt injection method (10:1 shunt ratio) using a 1.0 μL injection volume. Helium served as the carrier gas at a flow rate of 1.4 mL/min. For MS conditions, an EI source was employed with a temperature of 230 °C for the ion source and 280 °C for the interface. The ionization energy was set to 70 eV, and the scanning mode used for quantification was the ion scanning mode (SIM), with the full scanning mode (SCAN) used for qualitative reference. A solvent delay time of 4.3 min was applied. A total of 21 contaminants were detected in the surface water, primarily divided into three categories: 10 polycyclic aromatic hydrocarbons (PAHs), 8 organophosphates, and 3 organochlorines. In the sediments, 19 contaminants were detected, categorized into four groups: 3 polycyclic aromatic hydrocarbons, 3 pyrethroids, 8 organophosphate esters, and 5 organochlorines.

2.3. Quality Assurance/Quality Control

The QA/QC method mainly includes spiked blank, method blank, matrix spiked parallel, and sample parallel samples, and the whole experimental process refers to the QA/QC quality control system of the USEPA TO-13A method [27]. One liter each of ultrapure water and blank filter membrane were used as the field blank and experimental process blank samples, respectively, and the field blank, process blank, and collected samples were subjected to the same extraction, purification, and analytical detection processes to evaluate possible contamination during the whole experimental process. The spiked recoveries of pesticides in the water and sediment samples ranged from 60% to 140%. The method detection limits and method quantification limits are shown in Table S2.

2.4. Ecological Risk Assessment

Ecological risk assessments of pesticides in the Danjiangkou Reservoir were performed using the Hazard Quotient (HQ) method [28–30]. Data regarding the toxicity of pesticides to aquatic organisms were collected from the ECOTOX database (https://cfpub.epa.gov/ecotox/ (accessed on 16 October 2023)) and other published research (see Tables S4 and S5 for details). The HQ was calculated as follows:

$$HQ_{water} = EC_{water}/PNEC_{water} \quad (1)$$

$$PNEC_{water} = L(E)C_{50}/AF \quad (2)$$

where EC_{water} is the contaminant concentrations in water, ng/L. $PNEC_{water}$ is the predicted no effect concentration of contaminants in water, mg/L. AF is the evaluation factor, which is taken as 1000 according to the EU Water Framework Directive. $L(E)C_{50}$ is the median lethal concentration of the tested organisms. ECOSAR software 2.2 was used to simulate the toxicity data of the pesticides in relation to fish, daphnia, and green algae (Tables S4 and S5). When $HQ_{water} < 0.01$ and there is no detectable concentration, the ecological risk is insignificant. When $0.01 < HQ_{water} < 0.1$, the ecological risk is low. When $0.1 < HQ_{water} < 1$, the ecological risk is medium. When $HQ_{water} \geq 1$, the ecological risk is high.

According to EU technical guidelines on environmental risk assessment, $PNEC_{sed}$ (ng/g) was calculated by the equilibrium partitioning method.

$$PNEC_{sediment} = \frac{K_{susp-water} \times 1000 \times PNEC_{water}}{RHO_{susp}} \quad (3)$$

$$K_{susp-water} = F_{water-susp} + F_{solid-susp} \times F_{oc-susp} \times \frac{K_{oc}}{1000} \times RHO_{solid} \quad (4)$$

where $K_{\text{susp-water}}$ is the suspended matter-water partitioning coefficient (m^3/m^3). RHO_{susp} is the suspended solids capacity (1150 kg/m^3). RHO_{solid} is the density of the solid phase (2500 kg/m^3). $F_{\text{water-susp}}$ is the volume ratio of water in suspended matter (0.9 m^3/m^3). $F_{\text{solid-susp}}$ is the volume ratio of solids in suspended matter (0.1 m^3/m^3). $F_{\text{oc-susp}}$ is the mass fraction of organic carbon in suspended matter (0.1 kg/kg). K_{oc} represents the absorption constants of organic compounds (L/kg), which can be obtained by EPI suit v4.11.

$$HQ_{sed} = EC_{sed}/PNEC_{sed} \tag{5}$$

HQ is used to assess the ecological risk of contaminants in samples from the Danjiangkou Reservoir. EC_{sed} is the concentration of the contaminant in the sediment (ng/g). For compounds with log K_{ow} values less than 5, the quotient HQ_{sed} is compared to 1 (threshold). For compounds with log K_{ow} values greater than 5, the HQ_{sed} is compared to 10. This is because contaminants with log K_{ow} values greater than 5 have a higher capacity for sediment adsorption and are less harmful to benthic organisms, so the evaluation criteria can be relaxed [28].

2.5. Human Health Risk Assessment

To estimate human health risks associated with exposure to PAHs and pesticides, the U.S. Environmental Protection Agency's (EPA) health risk assessment model was employed to quantify both carcinogenic and non-carcinogenic risks. Exposure to these substances in water occurs through dermal contact and ingestion, typically during swimming and drinking. Similarly, exposure to sediments happens via ingestion of soil and dermal contact with it. To represent a worst-case exposure scenario, risk calculations utilized adult body weight values (70 kg). The methodology for integrating toxicity and exposure data in this study adhered to the risk assessment guidelines [31]. Assessments were conducted for risks arising from ingestion of surface water and sediment, as well as dermal contact with both.

$$\text{CDI(Water Ingestion)} = \frac{CW \times IR(W) \times EF \times ED}{BW \times AT}$$

CDI (Water Dermal) was calculated with reference to a previous study [32].

$$\text{CDI(Water Dermal)} = \frac{(6 \times \tau \times \frac{TE}{\pi})^{0.5} \times CW \times k \times SA \times EF \times FE \times ED}{BW \times AT \times f \times 500}$$

where CDI is the daily exposure of the *i*-th contamination (mg/kg·d). CW is the chemical concentration in water (mg/L). IR(W) is the ingestion rate. EF is the exposure frequency. ED is the exposure duration. BW is the body weight. AT is the averaging time. TE is the event duration. f is the intestinal absorption ratio. k is a skin permeability parameter. τ is the lag time for each pollutant in the body. FE is the event frequency. SA is the skin surface area available for contact. ET is the exposure time. CF is the volumetric conversion factor for water.

$$\text{CDI(Sediment Ingestion)} = \frac{CS \times IR(S) \times CF \times FI \times EF \times ED}{BW \times AT}$$

$$\text{CDI(Sediment Dermal)} = \frac{CS \times SA \times AF \times ABS \times EF \times ED \times CF}{BW \times AT}$$

where CS is the chemical concentration in soil (mg/kg). IR(S) is the ingestion rate (mg/d). CF is the conversion factor. FI is the fraction ingested from contaminated sources. EF is the exposure frequency. ED is the exposure duration. BW is the body weight. AT is the averaging time. SA is the skin surface area available for contact. AF is the soil to skin adherence factor. ABS is the absorption factor.

The non-carcinogenic risk assessment model is as follows [33]:

$$HQ = \frac{CDI}{RfD}$$

$$HI = \sum_{i=1}^{n} HQ_i$$

where RfD denotes the reference dose (mg/kg·d). HQ is the non-carcinogenic risk index. HI is the total non-carcinogenic risk index. In order to estimate the total health risk from complex pollutants (risk from exposure to different pollutants), the hazard index was applied. The HI was estimated as the sum of the HQs for the individual pollutants. The HI exceeded 1, which means that it presents a potentially adverse human health effect [34].

The carcinogenic risk assessment model is as follows:

$$CR_i = CDI_i \times SF_i$$

$$TCR = \sum_{i=1}^{n} CR_i$$

where SF is the carcinogenic slope factor (mg/kg·d). R_i is the carcinogenic risk index. TCR is the total carcinogenic risk index. For most regulatory programs, a carcinogenic risk value between 10^{-6} and 10^{-4} indicates a potential risk, while a carcinogenic risk value of 10^{-6} or less indicates a safe level, and a carcinogenic risk value greater than 10^{-4} indicates a potentially high health risk.

3. Results and Discussion

3.1. Occurrence and Spatial Distribution of Contaminants in Surface Water of Danjiangkou Reservoir

In water samples collected from the Danjiangkou Reservoir, a total of 21 pollutants were detected, including 10 PAHs and 11 pesticides (Table 1). As shown in Figure 2, it is evident that throughout the 23 monitoring points, polycyclic aromatic hydrocarbons (PAHs) dominate, with proportions ranging from 60.79% to 99.15%. Organic phosphorus pesticides vary in proportion from 0.72% to 38.90%, while organochlorine pesticides exhibit low detection concentrations across all measuring points, with proportions ranging from 0.00% to 2.22%. These results are consistent with previous conclusions drawn from surface water in the Yongding River basin [35].

Specifically, the total concentration of PAHs (∑PAHs) in the reservoir's water ranged from 64.64 to 868.23 ng/L, averaging at 217.97 ± 184.97 ng/L. In comparison to other Chinese rivers, such as the Qiantang River (70.3 to 1844.4 ng/L, average: 283.3 ng/L), the Yellow River (144.3 to 2360 ng/L, average: 662 ng/L), and the Daliao River Estuary and adjacent areas (71.12 to 4255.43 ng/L, average: 748.76 ng/L), the ∑PAH concentrations in the Danjiangkou Reservoir were relatively lower [36–38], and the concentration of ∑PAHs in the Danjiangkou Reservoir was also low compared to that in foreign countries, such as in the Mahanadi River Basin (13.1–685.4 μg/L) and on the Mediterranean coast (13,156–34,852 ng/g) [39,40]. However, a comparison of the concentrations of specific pollutants revealed that both Nap and Phe were the pollutants with the highest detected concentrations. As shown in Figure 2, the concentration of ∑PAHs varied among the tributary sampling sites, and among all sampling sites, the highest concentration was found at the Guanshan River (R3). This may be attributed to the intensive human activity and fossil fuel combustion in nearby areas [41]. In contrast, the concentration of ∑PAHs presented small fluctuations among the sampling sites in the Han and Dan Reservoirs, resulting in higher average concentrations than some tributary rivers (i.e., Jianhe River and Lang River) with less human activity. The sources of PAHs in the Han and Dan Reservoirs may include the input from tributary rivers as well as ship operations and oil leakage [42]. Notably, PAHs containing 2–3 benzene rings and 4 benzene rings (low molecular weight PAHs, LMW-PAHs) were predominant in surface water from the Danjiangkou Reservoir, with naphthalene as the most predominant. This is similar to findings in the Three Gorges Reservoir [43]. Naphthalene is a major product of diesel and gasoline, possibly originating from incomplete combustion sources, while pyrene is indicative of coal combustion and benzo[a]pyrene is a marker for fuel combustion. Therefore, it can be inferred that the primary sources of PAHs in the study area are petroleum pollution and combustion of petroleum products [44].

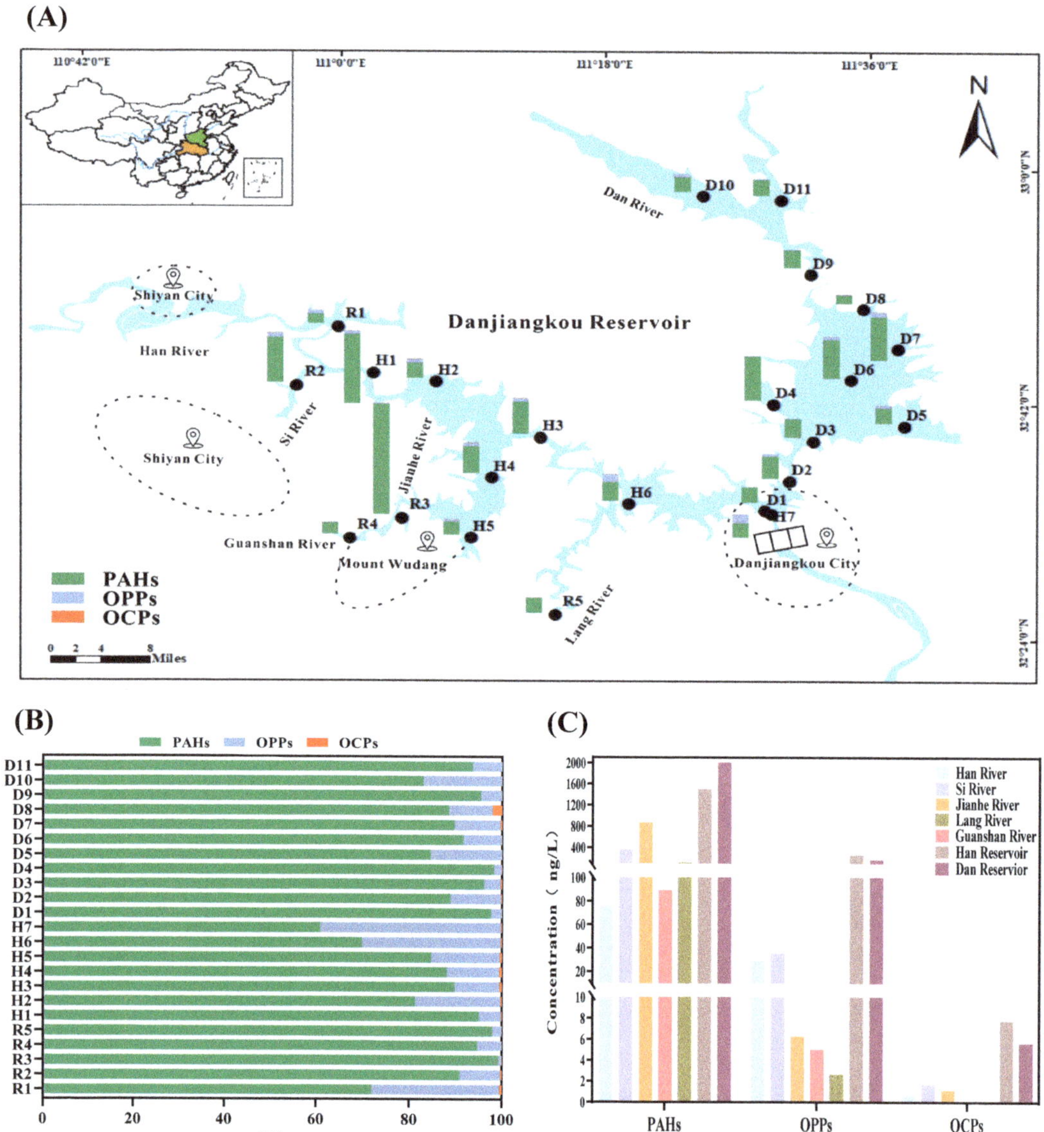

Figure 2. The occurrence of three types of organic pollutants in surface waters of the Danjiangkou Reservoir. (**A**) The spatial distributions of three types of organic pollutants in the Danjiangkou Reservoir. (**B**) The composition of three types of organic pollutants in surface water of the Danjiangkou Reservoir. (**C**) A histogram of the organic pollutant concentrations in different areas of the Danjiangkou Reservoir. PAHs: polycyclic aromatic hydrocarbons; OPPs: organophosphates; OCPs: organochlorines.

Table 1. The detection results of PAHs, organophosphates, and organochlorine pesticides and their isomers in water samples from the Danjiangkou Reservoir.

Targets	Analyte	Abbreviation	DF (%)	Water (n = 29) ng/L Min	Max	Mean
PAHs	Naphthalene	Nap	100.0	10.43	116.97	41.06
	Acenaphthylene	Acy	-	ND	ND	-
	Acenaphthene	Ace	4.4	ND	0.22	0.11
	Fluorene	Flu	100.0	22.74	87.61	42.45
	Phenanthrene	Phe	100.0	26.54	162.86	71.43
	Fluoranthene	Fla	8.7	ND	7.56	5.64
	Pyrene	Pyr	4.4	ND	12.04	12.04
	Benz[a]anthracene	BaA	4.4	ND	13.12	13.12
	Benzo[a]pyrene	BaP	13.0	ND	66.40	28.61
	Indeno[1,2,3-cd]pyrene	InP	17.4	ND	750.42	331.85
OPPs	O,O,O-triethyl-Phosphorothioate	O,O,O-Tept	-	ND	ND	-
	Dichlorvos	DDVP	78.3	ND	3.15	2.57
	E-Mevinphos	TRANS	60.9	ND	9.56	9.23
	Tetraethyl pyrophosphate	TEPP	4.4	ND	38.74	38.74
	Naled	BRP	4.4	ND	40.00	40.00
	Phorate	Phr	4.4	ND	2.39	2.39
	Simazine	DCT	-	ND	ND	-
	Atrazine	ATZ	91.3	ND	29.03	12.29
OCPs	α-HCH	α-HCH	95.7	ND	40.96	0.61
	β-HCH	β-HCH	43.5	ND	97.73	1.09
	γ-HCH	γ-HCH	65.2	ND	0.72	1.11

DF: Detection frequency. ND: Means a concentration below the method detection limit.

The total concentrations of organophosphorus pesticides (∑OPPs) in surface water from the Danjiangkou Reservoir ranged from 2.62 to 71.14 ng/L, averaging at 22.26 ± 18.06 ng/L, which were much lower than those in the Zhujiang Estuary (∑OPPs: 460–43,600 ng/L) and in the water samples of Jiangsu Province (∑OPPs: 63.7–22,463 ng/L) [45], and the concentrations in surface water samples from foreign countries were lower than those of the Tiber River (∑OPPs: 0.40–224.48 ng/L) [46]. Among the detected OPPs, atrazine and dichlorvos had higher detection rates and average concentrations, which is consistent with the findings of previous studies [47]. Spatially, the concentrations of ∑OPPs were higher in the Han Reservoir and Dan Reservoir than in the tributary sampling sites. Additionally, the Han River (R1) and Si River (R2) had higher concentrations of ∑OPPs than other tributaries, possibly due to higher levels of agricultural activity in these locations [48].

Furthermore, the study revealed that the total concentration of organochlorine pesticides (∑OCPs) in surface water within the research area varied from non-detectable levels to 1.75 ng/L, averaging at 0.73 ± 0.66 ng/L. According to previous studies, low levels of organochlorine pesticides (OCPs) were detected in the Xujia River Basin, the Beiluo River, and the Irtysh River, which were similar to those found in the surface water of the Danjiangkou Reservoir, with high detection rates of α-HCH, β-HCH, and γ-HCH [49–51]. The lower concentrations of organochlorine pesticides compared with organophosphorus pesticides may be related to the restriction of the production and use of certain OCPs since 1983 and the ratification of the Stockholm Convention on Persistent Organic Pollutants (POPs) [52]. The lower concentrations of ∑OCPs at the tributary sampling sites also indicate limited input from neighboring regions, and the relatively higher concentrations in the Han Reservoir and Dan Reservoir may be related to the extensive historical use and persistence of organochlorine pesticides [3,53,54]. Among the detected OCPs, HCH-alpha had the highest detection frequency and concentration, which may primarily stem from residual organochlorine pesticide usage in the past [55].

Taken together, the total concentrations of detected pollutants ranged from nondetectable levels (ND) to several thousand nanograms per liter (ng/L), aligning with findings from global studies [28,56,57]. Our results also indicate that polycyclic aromatic hydrocarbons (PAHs), specifically Nap, Flu, and Phe, are the primary pollutants in the surface water of the Danjiangkou Reservoir. The Han Reservoir and Dan Reservoir had higher concentrations of all the three types of detected pollutants (PAHs, OPPs, and OCPs) compared to the tributaries, suggesting that the reservoirs may have a sedimentation and aggregation effect on the surrounding input pollutants.

3.2. Occurrence and Spatial Distribution of Contaminants in Sediments of Danjiangkou Reservoir

In sediment samples collected from the Danjiangkou Reservoir, a total of 19 pollutants were analyzed, including 3 PAHs, 8 organophosphates, and 5 organochlorines (Table 2 and Figure 3). The total concentration of PAHs in the sediment samples ranged from 0.01 to 2.93 ng/g (average: 0.69 ± 0.94 ng/g), significantly lower than the PAH concentrations found in the Yangtze River Wuhan segment (72.4–3995.2 ng/g) and the Yangtze River Delta urban agglomeration (493–15,907 ng/g; average: 2483 ng/g) [58,59]. Furthermore, the PAH concentrations in the Danjiangkou Reservoir sediment were much lower than those in other Chinese reservoirs like the Chahe Reservoir (642.94–1495.96 ng/g) and the Fenhe Reservoir and its basin (539.0–6281.7 ng/g; average: 2214.8 ng/g) [60,61]. The Mahanadi River, India (13.1–685.4 ng/g), and the Tiber River, Italy (4.5–652.2 ng/g), also had much higher concentrations than those measured in sediments from the Danjiangkou Reservoir [62]. These results suggest a relatively minimal pollution level of PAHs in the Danjiangkou Reservoir. In contrast with the above results of the surface water samples, the detection rates of high molecular weight (HMW) PAHs were all above 60%, whereas LMW PAHs were not detected in the sediment samples. This may be because HMW PAHs (4–6 ring PAHs) have stronger oil affinity, lower volatility, and lower water solubility, and are more prone to accumulate in sediments, potentially posing carcinogenic effects in humans [63–65].

Unlike the surface water results, organophosphate pesticides (OPPs) were dominant in the sediment at most sampling sites. The total concentration of ∑OPPs in the reservoir sediment ranged from nd-25.57 ng/g (average: 6.93 ± 8.41 ng/g), which is much lower than that reported in the Huai River basin (2951–47,739 ng/g; average: 8041 ng/g) and the Yangtze River basin (3022–12,756 ng/g; average: 5936 ng/g) [45]. Among the detected OPPs, the highest detection frequency of 65.22% was observed for TRANS, indicating its more prevalent use in adjacent areas. Additionally, the occurrence of pyrethroids was detected in the sediment samples from the Danjiangkou Reservoir, with the total concentrations ranging from nd-1.78 ng/g (average: 0.67 ± 0.60 ng/g). These were lower than the concentrations found in sediment samples from urban rivers in Guangzhou in both the dry season (0.25–3113 ng/g; average: 217 ng/g) and the wet season (0.25–4732 ng/g; average: 388 ng/g) [66]. The pyrethroid concentrations in the Danjiangkou Reservoir sediment were also lower than those in sediments from the Liao River basin (2.2–102.5 ng/g; average: 17.3 ng/g) [48]. Furthermore, the total concentrations of organochlorine pesticides (OCPs) in the sediment samples from the Danjiangkou Reservoir (nd-1.11 ng/g, average: 0.39 ± 0.27 ng/g) were comparable to the reported concentrations in sediments from the Dongping Lake in North China and the Chennab River in Pakistan [67,68], but much lower than those detected in the Huai River basin (531–6533 ng/g; average: 1408 ng/g) and the Yangtze River basin (305–2190 ng/g; average: 1077 ng/g) [45]. Among the detected OCPs, beta-hexachlorocyclohexane (β-HCH) had the highest detection rate (exceeding 80%) and dieldrin had the highest concentration (up to 0.98 ng/g). These results indicate relatively low environmental concentrations in the sediment of the Danjiangkou Reservoir.

Taken together, the components of the detected pollutants in the sediment varied from those in the surface water of the Danjiangkou Reservoir. For one thing, the OPPs were dominant at most sampling sites, and the occurrence of pyrethroids was detected in the sediment. For another, HMW PAHs were the primary PAHs in the sediment, and no LMW PAHs were detected. However, the average concentrations of all three types of detected pollutants (PAHs, OPPs, and OCPs) were also higher in the Han Reservoir and

Dan Reservoir than the tributaries, which is consistent with the findings in surface water. These results also confirm the speculation that the reservoirs have a sedimentation and aggregation effect on the surrounding input pollutants.

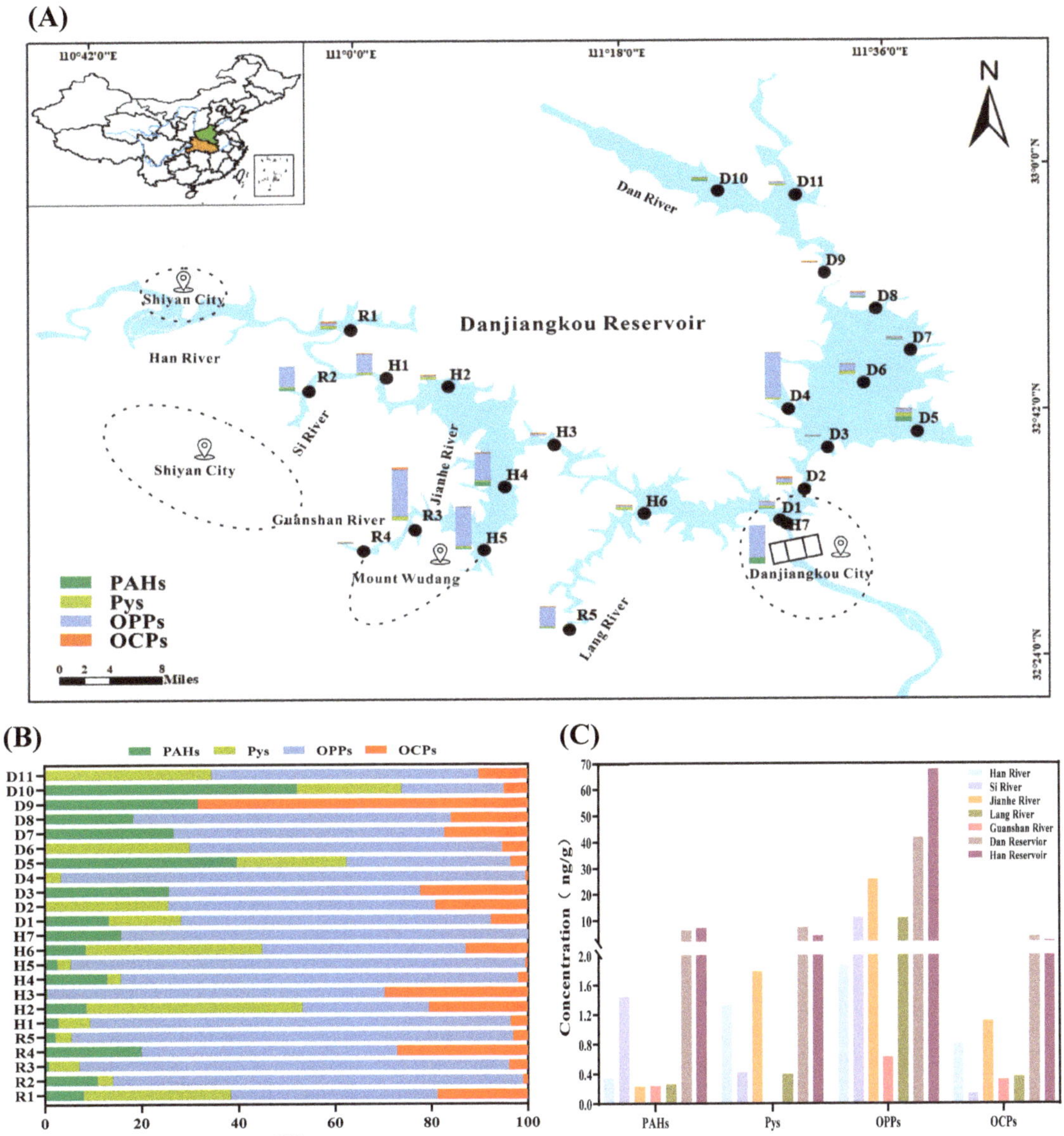

Figure 3. The occurrence of four types of organic pollutants in sediment of the Danjiangkou Reservoir. (**A**) The spatial distributions of four types of organic pollutants in the Danjiangkou Reservoir. (**B**) The composition of four types of organic pollutants in sediment of the Danjiangkou Reservoir. (**C**) A histogram of organic pollutant concentrations in different areas of the Danjiangkou Reservoir. PAHs: polycyclic aromatic hydrocarbons; Pys: pyrethroids; OPPs: organophosphates; OCPs: organochlorines.

Table 2. The detection results of PAHs, pyrethroids, organophosphates, and organochlorine pesticides and their isomers in sediments from the Danjiangkou Reservoir.

Targets	Analyte	Abbreviation	DF (%)	Sediment (n = 23) ng/g Min	Max	Mean
PAHs	Benz[a]anthracene	BaA	100.0	0.00313	0.01	0.01
	Benzo[k]fluoranthene	BkF	60.9	ND	2.87	0.77
	Indeno[1,2,3-cd]pyrene	InP	69.6	ND	0.75	0.31
Pys	Cyhalothrin (Lambda)	Cya	56.5	ND	1.73	0.65
	Cypermethrin I	Cye	26.1	ND	0.72	0.47
	Fenvalerate I	Fev	21.7	ND	1.78	0.82
OPPs	Dichlorvos	DDVP	17.4	ND	0.85	0.31
	E-Mevinphos	TRANS	65.2	ND	0.72	0.54
	Simazine	DCT	8.7	ND	2.23	1.38
	Malathion	Maa	43.5	ND	9.16	2.61
	Prothiofos	Prt	43.5	ND	1.46	0.66
	Sulprofos	Sup	-	ND	ND	-
	Famphur	Fap	13.0	ND	1.53	1.18
	Azinphos-methyl	Azn	30.4	ND	21.38	15.87
OCPs	β-HCH	β-HCH	87.0	ND	0.87	0.26
	Pentachloronitrobenzene	Pet	52.2	ND	0.13	0.10
	δ-HCH	δ-HCH	30.4	ND	0.21	0.17
	Chlordane-trans	Cht	13.0	ND	0.16	0.10
	Dieldrin	Dil	4.4	ND	0.98	0.98

DF: Detection frequency. ND: Means a concentration below the method detection limit.

3.3. Ecological Risk Assessment of Danjiangkou Reservoir

During this study, we conducted an ecological risk assessment of the Danjiangkou Reservoir by examining the potential risks for three trophic levels: green algae, water fleas, and fish. The hazard quotient (HQ) was employed as an effective method to evaluate risks based on measured environmental concentrations [28]. An HQ value greater than 1 indicates high ecological risk, between 0.1 and 1 signifies moderate risk, less than 0.1 indicates low risk, and values below 0.01 or undetectable pollutant concentrations are considered insignificant.

Among the contaminants detected in surface water from the Danjiangkou Reservoir, naphthalene, fluorene, and phenanthrene may pose high risks to green algae and fish at the H4, H7, and D6 locations, and Phe may pose medium to high risks to water fleas and fish at all sampling sites (Figure 4). At H4, BaA and BaP were found to pose high risks to green algae and fish. InP posed medium to high risks to all three species at the R2, R5, H1, and D4 locations. Additionally, organophosphate pesticides such as DDVP, TRANS, BRP, Phr, and ATZ posed high risks to green algae and fish at certain points. BRP and Phr posed high risks to water fleas at H7. Among the three organochlorine pesticides, α-HCH showed medium risks for green algae and fish at three identical points, while Lindane (γ-HCH) posed high risks for green algae and fish at the same location. Taken together, PAHs such as phenanthrene may pose higher risks to water fleas and fish compared to other detected contaminants, making them priority pollutants that require further attention. In addition, aquatic organisms living around sampling sites H4, H7, and D6 may encounter moderate to high risks due to the presence of various pollutants, necessitating closer scrutiny of emissions from nearby pollution sources.

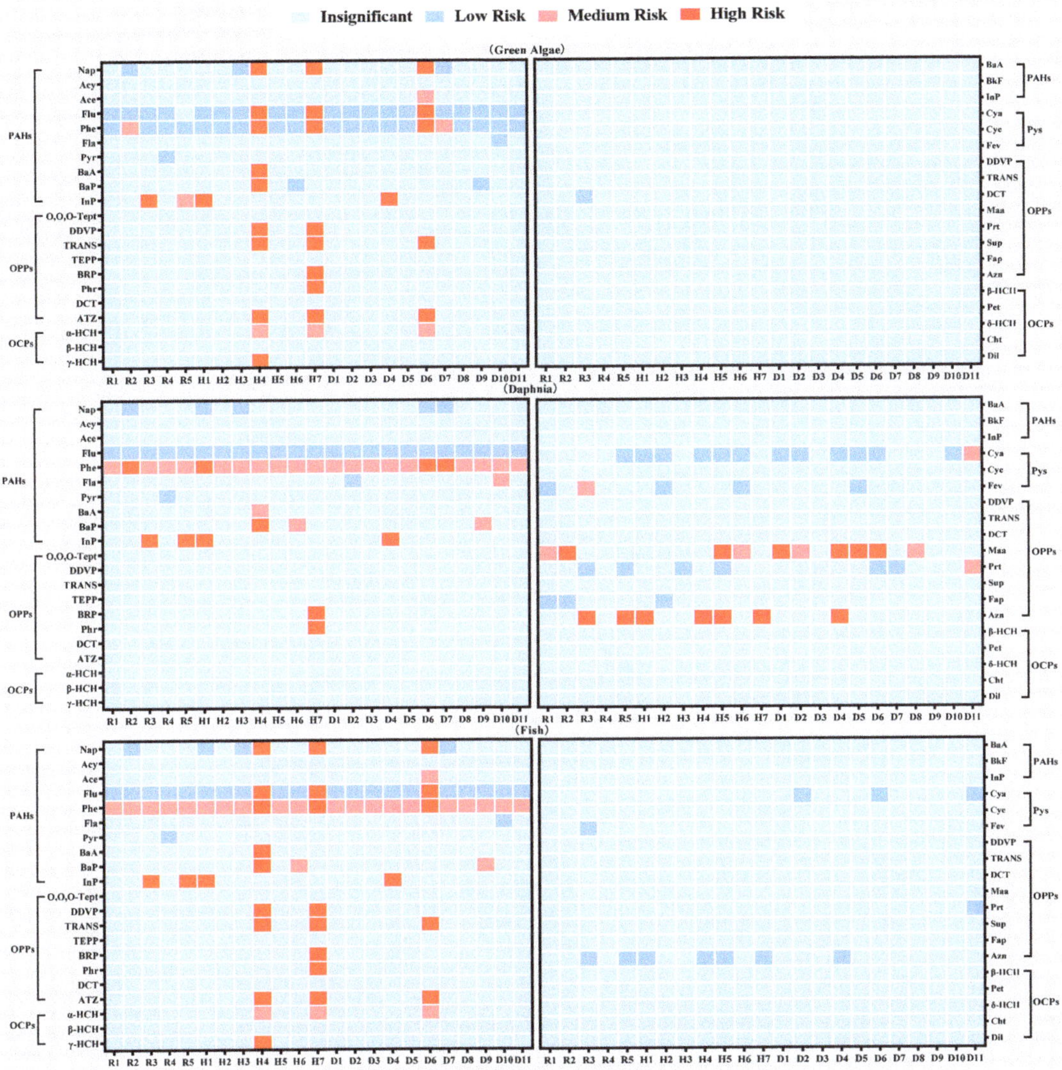

Figure 4. A heatmap of the risk quotients of the organic pollutants detected at each sampling point for the three most sensitive species. Hazard quotients (HQs) are based on the concentrations of organic pollutants measured in surface water and sediment at each sampling point in the Danjiangkou Reservoir. Nap: Naphthalene; Acy: Acenaphthylene; Ace: Acenaphthene; Flu: Fluorene; Phe: Phenanthrene; Fla: Fluoranthene; Pyr: Pyrene; BaA: Benz[a]anthracene; BaP: Benzo[a]pyrene; InP: Indeno[1,2,3-cd]pyrene; O,O,O-Tept: O,O,O-triethyl-Phosphorothioate; DDVP: Dichlorvos; TRANS: E-Mevinphos; TEPP: Tetraethyl pyrophosphate; BRP: Naled; Phr: Phorate; DCT: Simazine; ATZ: Atrazine; BkF: Benzo[k]fluoranthene; Cya: Cyhalothrin (Lambda); Cye: Cypermethrin I; Fev: Fenvalerate I; DCT: Simazine; Maa: Malathion; Prt: Prothiofos; Sup: Sulprofos; Fap: Famphur; Azn: Azinphos-methyl; Pet: Pentachloronitrobenzene; Cht: Chlordane-trans; Dil: Dieldrin.

To evaluate the ecological risks posed by contaminants detected in the sediments from the Danjiangkou Reservoir, based on the guidelines recommended by the European Union, the equilibrium partitioning method was utilized to calculate the predicted non-effect concentration [28]. As shown in Figure 4, neither PAHs nor organochlorine pesticides in sediment would pose potential risks to the three categories of organisms, whereas Cya, Fev, and Azn may pose low risks to fish at specific monitoring sites. However, low to medium risks for water fleas were observed in three pyrethroid pesticides at certain sites (R5 to D11). In contrast, the organophosphate pesticides Maa and Azn may pose a high level of risk to water fleas at certain sites (Maa: R2, H5, D1, D4, D5, and D6; Azn: R3, R5, H1, H4, H5, H7, and D4). This may be related to their higher concentrations in sediments as abovementioned in the chemical analysis results.

Overall, our results suggest that water fleas are more sensitive to the contaminants in sediment of the reservoir, especially Maa and Azn. These observations also indicate that organophosphate pesticides might be the primary risk sources in the sediment, which should be considered a priority.

3.4. Human Health Risk Assessment of Danjiangkou Reservoir

The non-carcinogenic and carcinogenic risks were calculated following USEPA protocols to evaluate whether the detected contaminants in surface water could cause any potential threat to human health [69]. Table 3, Tables S8 and S10 list the HIs obtained by the non-carcinogenic risk evaluation of the contaminants in surface water and sediment at different sampling sites in the Danjiangkou Reservoir, and all the calculated HIs obtained were less than 1, which indicates that these contaminants may not have negative health effects [70]. Table 4, Tables S9 and S11 show the TCRs obtained from the carcinogenic risk evaluation of the pollutants in surface water and sediments in the study area, and the calculated values were all below 10^{-6}, indicating that these contaminants may not pose a carcinogenic risk.

Table 3. Health risks of typical pollutants in surface water of Danjiangkou Reservoir.

Site	Non-Carcinogenic Risks of Surface Water						Carcinogenic Risks of Surface Water		
	Naphthalene	Fluorene	Phenanthrene	Dichlorvos	Atrazine	α-HCH	Naphthalene	Dichlorvos	α-HCH
R1	3.02×10^{-5}	1.79×10^{-5}	2.82×10^{-5}	1.37×10^{-4}	1.44×10^{-5}	2.41×10^{-6}	7.25×10^{-8}	1.98×10^{-8}	1.22×10^{-7}
R2	1.67×10^{-4}	5.74×10^{-5}	1.55×10^{-4}	0	2.12×10^{-5}	1.99×10^{-6}	4.01×10^{-7}	0	1.00×10^{-7}
R3	2.66×10^{-5}	3.39×10^{-5}	4.92×10^{-5}	1.44×10^{-4}	3.06×10^{-6}	0	6.38×10^{-8}	2.09×10^{-8}	0
R4	3.27×10^{-5}	1.86×10^{-5}	2.67×10^{-5}	1.56×10^{-4}	1.88×10^{-6}	0	7.84×10^{-8}	2.26×10^{-8}	0
R5	2.66×10^{-5}	1.62×10^{-5}	2.53×10^{-5}	1.50×10^{-4}	0	0	6.39×10^{-8}	2.17×10^{-8}	0
H1	9.64×10^{-5}	4.18×10^{-5}	1.28×10^{-4}	1.21×10^{-4}	1.34×10^{-5}	2.03×10^{-6}	2.31×10^{-7}	1.76×10^{-8}	1.02×10^{-7}
H2	4.40×10^{-5}	2.01×10^{-5}	6.04×10^{-5}	1.40×10^{-4}	1.31×10^{-5}	2.50×10^{-6}	1.06×10^{-7}	2.03×10^{-8}	1.26×10^{-7}
H3	1.25×10^{-4}	4.57×10^{-5}	9.47×10^{-5}	0	1.47×10^{-5}	2.47×10^{-6}	3.00×10^{-7}	0	1.25×10^{-7}
H4	4.40×10^{-5}	2.01×10^{-5}	7.30×10^{-5}	1.43×10^{-4}	1.30×10^{-5}	2.25×10^{-6}	1.06×10^{-7}	2.07×10^{-8}	1.13×10^{-7}
H5	3.13×10^{-5}	1.89×10^{-5}	5.31×10^{-5}	1.25×10^{-4}	5.71×10^{-6}	2.38×10^{-6}	7.51×10^{-8}	1.81×10^{-8}	1.20×10^{-7}
H6	4.72×10^{-5}	2.51×10^{-5}	6.31×10^{-5}	1.58×10^{-4}	9.61×10^{-6}	2.56×10^{-6}	1.13×10^{-7}	2.29×10^{-8}	1.29×10^{-7}
H7	4.99×10^{-5}	2.47×10^{-5}	3.97×10^{-5}	1.34×10^{-4}	1.41×10^{-5}	2.04×10^{-6}	1.20×10^{-7}	1.95×10^{-8}	1.03×10^{-7}
D1	3.20×10^{-5}	3.05×10^{-5}	5.49×10^{-5}	1.80×10^{-4}	0	0	7.69×10^{-8}	2.61×10^{-8}	0
D2	5.24×10^{-5}	3.60×10^{-5}	8.13×10^{-5}	0	1.02×10^{-5}	2.06×10^{-6}	1.26×10^{-7}	0	1.04×10^{-7}
D3	2.82×10^{-5}	3.76×10^{-5}	6.92×10^{-5}	1.54×10^{-4}	2.24×10^{-6}	2.23×10^{-6}	6.76×10^{-8}	2.23×10^{-8}	1.12×10^{-7}
D4	4.01×10^{-5}	2.75×10^{-5}	3.77×10^{-5}	1.47×10^{-4}	2.87×10^{-6}	1.90×10^{-6}	9.63×10^{-8}	2.14×10^{-8}	9.56×10^{-8}
D5	4.25×10^{-5}	2.80×10^{-5}	5.00×10^{-5}	0	1.07×10^{-5}	0	1.02×10^{-7}	0	0
D6	1.49×10^{-4}	4.72×10^{-5}	1.26×10^{-4}	0	1.50×10^{-5}	2.00×10^{-6}	3.59×10^{-7}	0	1.01×10^{-7}
D7	1.51×10^{-4}	6.26×10^{-5}	1.45×10^{-4}	0	2.37×10^{-5}	2.20×10^{-6}	3.61×10^{-7}	0	1.11×10^{-7}
D8	1.49×10^{-5}	1.92×10^{-5}	2.61×10^{-5}	1.39×10^{-4}	3.61×10^{-6}	1.80×10^{-6}	3.58×10^{-8}	2.02×10^{-8}	9.06×10^{-8}
D9	2.88×10^{-5}	2.31×10^{-5}	7.58×10^{-5}	1.41×10^{-4}	3.66×10^{-6}	0	6.92×10^{-8}	2.05×10^{-8}	0
D10	2.41×10^{-5}	2.23×10^{-5}	5.97×10^{-5}	1.54×10^{-4}	1.02×10^{-5}	0	5.79×10^{-8}	2.24×10^{-8}	0
D11	6.48×10^{-5}	2.29×10^{-5}	4.26×10^{-5}	1.68×10^{-4}	4.48×10^{-6}	0	1.56×10^{-7}	2.43×10^{-8}	0

Table 4. Health risks of typical pollutants in sediment of Danjiangkou Reservoir.

Site	Non-Carcinogenic Risks of Sediment					Carcinogenic Risks of Sediment			
	Cyhalothrin (Lambda)	**Fenvalerate I**	**Dichlorvos**	**Azinphos-Methyl**	**Dieldrin**	**Benz[a]Anth-Racene**	**Benzo[k]flu-Oranthene**	**Indeno[1,2,3-cd]Pyrene**	**β-HCH**
R1	3.02×10^{-5}	1.79×10^{-5}	2.82×10^{-5}	1.37×10^{-4}	1.44×10^{-5}	1.38×10^{-11}	1.70×10^{-12}	3.65×10^{-10}	1.36×10^{-9}
R2	1.67×10^{-4}	5.74×10^{-5}	1.55×10^{-4}	0	2.12×10^{-5}	1.37×10^{-11}	2.27×10^{-11}	0	0
R3	2.66×10^{-5}	3.39×10^{-5}	4.92×10^{-5}	1.44×10^{-4}	3.06×10^{-6}	1.79×10^{-11}	1.36×10^{-12}	3.72×10^{-10}	1.24×10^{-9}
R4	3.27×10^{-5}	1.86×10^{-5}	2.67×10^{-5}	1.56×10^{-4}	1.88×10^{-6}	1.07×10^{-11}	0	3.79×10^{-10}	1.17×10^{-9}
R5	2.66×10^{-5}	1.62×10^{-5}	2.53×10^{-5}	1.50×10^{-4}	0	1.33×10^{-11}	3.49×10^{-12}	0	0
H1	9.64×10^{-5}	4.18×10^{-5}	1.28×10^{-4}	1.21×10^{-4}	1.34×10^{-5}	1.01×10^{-11}	0	3.63×10^{-10}	8.16×10^{-10}
H2	4.40×10^{-5}	2.01×10^{-5}	6.04×10^{-5}	1.40×10^{-4}	1.31×10^{-5}	1.17×10^{-11}	0	0	1.39×10^{-9}
H3	1.25×10^{-4}	4.57×10^{-5}	9.47×10^{-5}	0	1.47×10^{-5}	1.48×10^{-11}	2.57×10^{-11}	1.18×10^{-9}	4.73×10^{-10}
H4	4.40×10^{-5}	2.01×10^{-5}	7.30×10^{-5}	1.43×10^{-4}	1.30×10^{-5}	1.54×10^{-11}	5.78×10^{-12}	3.55×10^{-10}	4.66×10^{-10}
H5	3.13×10^{-5}	1.89×10^{-5}	5.31×10^{-5}	1.25×10^{-4}	5.71×10^{-6}	1.03×10^{-11}	0	3.72×10^{-10}	1.44×10^{-9}
H6	4.72×10^{-5}	2.51×10^{-5}	6.31×10^{-5}	1.58×10^{-4}	9.61×10^{-6}	1.33×10^{-11}	4.81×10^{-13}	3.48×10^{-10}	7.40×10^{-10}
H7	4.99×10^{-5}	2.47×10^{-5}	3.97×10^{-5}	1.34×10^{-4}	1.41×10^{-5}	1.80×10^{-11}	4.55×10^{-11}	6.19×10^{-10}	0
D1	3.20×10^{-5}	3.05×10^{-5}	5.49×10^{-5}	1.80×10^{-4}	0	1.19×10^{-11}	4.32×10^{-12}	4.26×10^{-10}	5.45×10^{-10}
D2	5.24×10^{-5}	3.60×10^{-5}	8.13×10^{-5}	0	1.02×10^{-5}	1.61×10^{-11}	0	0	3.41×10^{-9}
D3	2.82×10^{-5}	3.76×10^{-5}	6.92×10^{-5}	1.54×10^{-4}	2.24×10^{-6}	1.47×10^{-11}	0	3.53×10^{-10}	4.94×10^{-10}
D4	4.01×10^{-5}	2.75×10^{-5}	3.77×10^{-5}	1.47×10^{-4}	2.87×10^{-6}	4.96×10^{-12}	4.85×10^{-13}	0	5.53×10^{-10}
D5	4.25×10^{-5}	2.80×10^{-5}	5.00×10^{-5}	0	1.07×10^{-5}	9.47×10^{-12}	4.02×10^{-11}	6.08×10^{-10}	3.53×10^{-10}
D6	1.49×10^{-4}	4.72×10^{-5}	1.26×10^{-4}	0	1.50×10^{-5}	1.06×10^{-11}	0	0	3.69×10^{-10}
D7	1.51×10^{-4}	6.26×10^{-5}	1.45×10^{-4}	0	2.37×10^{-5}	1.11×10^{-11}	3.45×10^{-12}	4.11×10^{-10}	8.96×10^{-10}
D8	1.49×10^{-5}	1.92×10^{-5}	2.61×10^{-5}	1.39×10^{-4}	3.61×10^{-6}	1.35×10^{-11}	5.41×10^{-12}	4.43×10^{-10}	1.83×10^{-9}
D9	2.88×10^{-5}	2.31×10^{-5}	7.58×10^{-5}	1.41×10^{-4}	3.66×10^{-6}	1.33×10^{-11}	0	3.97×10^{-10}	1.57×10^{-9}
D10	2.41×10^{-5}	2.23×10^{-5}	5.97×10^{-5}	1.54×10^{-4}	1.02×10^{-5}	1.38×10^{-11}	1.10×10^{-11}	7.80×10^{-10}	4.45×10^{-10}
D11	6.48×10^{-5}	2.29×10^{-5}	4.26×10^{-5}	1.68×10^{-4}	4.48×10^{-6}	8.09×10^{-12}	0	0	9.46×10^{-10}

It should be noted that the risk assessment in this study has certain limitations: (1) criteria derivation uncertainty. During the process of deriving PNECsediment, this study adopted the equilibrium distribution model based on some theoretical assumptions, and the differences between these assumptions and the real situation will certainly bring some uncertainty to the evaluation results; (2) risk assessment uncertainty. During the risk assessment based on the HQ method, the HQ is constructed based on the small amount of existing sediment toxicity data. A small amount of sediment toxicity data may cause uncertainty in the evaluation results. However, this study conducted risk assessments based on the detection concentrations of field samples from the water source area. In fact, the water still needs to undergo purification in a water treatment plant before it can be supplied for human consumption. Therefore, we can consider the Danjiangkou Reservoir as a safe water source.

4. Conclusions

This study primarily monitored the surface water and sediments of the Danjiangkou Reservoir, revealing the presence of various organic pollutants in the aquatic environment, including polycyclic aromatic hydrocarbons (PAHs) and pesticides. The concentration ranges detected in the surface water samples were as follows: ∑PAHs (64.64–868.23 ng/L, average concentration 217.97 ± 184.97 ng/L), ∑OPPs (2.62–71.14 ng/L, average concentration 22.26 ± 18.06 ng/L), and ∑OCPs (nd-1.75 ng/L, average concentration 0.73 ± 0.66 ng/L). In the sediments, the concentration ranges were as follows: ∑PAHs (0.01–2.93 ng/g, average concentration 0.69 ± 0.94 ng/g), ∑OPPs (nd-25.57 ng/g, average concentration 6.93 ± 8.41 ng/g), ∑Pys (nd-1.78 ng/g, average concentration 0.67 ± 0.60 ng/g), and ∑OCPs (nd-1.11 ng/g, average concentration 0.39 ± 0.27 ng/g). Although the concentrations of these pollutants are relatively low and show no significant differences compared to reports from other national and international river basins, assessments using the health risk evaluation model of the U.S. Environmental Protection Agency (EPA) indicate that they currently pose minimal risks to human health. However, ecological risk assessments conducted using the hazard quotients reveal potential risks posed by these pollutants to aquatic organisms, especially certain polycyclic aromatic hydrocarbons (such as phenanthrene) and organophosphate pesticides (such as malathion and chlorpyrifos). Therefore, continued monitoring and risk assessment are necessary to ensure the protection of aquatic ecosystems in the Danjiangkou Reservoir.

Supplementary Materials: The following supporting information can be downloaded at: https://www.mdpi.com/article/10.3390/toxics12120859/s1, Table S1: Danjiangkou Reservoir Points Information; Table S2: Recovery, method detection limit (MDL) and method quantification limit (MQL) of pesticides in surface water; Table S3: Recovery, method detection limit (MDL) and method quantification limit (MQL) of pesticides in sediment; Table S4: Aquatic toxicity data and Predicted No Effect Concentration (PNEC) values for polycyclic aromatic hydrocarbons (PAHs) and organophosphorus (OPs) and organochlorine (OC) pesticides in surface water samples; Table S5: Aquatic biotoxicity data for organophosphorus (OP), organochlorine (OC), and pyrethroid pesticides, and polycyclic aromatic hydrocarbons (PAH) in sediment samples, and logarithmic absorption constants (K_{oc}) and log K_{ow} for organic compounds; Table S6: Exposure parameters used for estimation of human health risk; Table S7: Reference dose (RfD) and slope factor (SF) for each pollutant; Table S8: Evaluation of non-carcinogenic risks of surface water pollutants; Table S9: Evaluation of carcinogenic risks of surface water pollutants; Table S10: Evaluation of non-carcinogenic risks of sediment pollutants; Table S11: Evaluation of carcinogenic risks of sediment pollutants.

Author Contributions: R.L.: project administration, funding acquisition, writing—review and editing; H.P.: writing—original draft, writing—review and editing; Y.G.: investigation; X.Z.: visualization; K.F.: data curation; T.Z.: software, formal analysis; L.Y.: methodology, writing—review and editing; J.H.: validation; B.Z.: supervision, writing—review and editing; S.Z.: resources, data curation. All authors have read and agreed to the published version of the manuscript.

Funding: This study was supported by the National Key Research and Development Program of China [grant number 2021YFC3200105].

Institutional Review Board Statement: Not applicable.

Informed Consent Statement: Not applicable.

Data Availability Statement: The raw data supporting the conclusions of this article will be made available by the authors on request.

Acknowledgments: We would like to thank the Analysis and Testing Center of the Institute of Hydrobiology for aiding with instrument analysis.

Conflicts of Interest: The authors declare no conflicts of interest.

References

1. Li, S.; Shi, W.; You, M.; Zhang, R.; Kuang, Y.; Dang, C.; Sun, W.; Zhou, Y.; Wang, W.; Ni, J. Antibiotics in water and sediments of Danjiangkou Reservoir, China: Spatiotemporal distribution and indicator screening. *Environ. Pollut.* **2019**, *246*, 435–442. [CrossRef] [PubMed]
2. Zheng, T.; Dang, C.; Zhong, S.; Sun, W.; Chen, Q. Spatiotemporal distribution, risk assessment and source appointment of metal(loid)s in water and sediments of Danjiangkou Reservoir, China. *Environ. Geochem. Health* **2021**, *43*, 139–152. [CrossRef] [PubMed]
3. Zhou, J.; Zheng, D.; Jin, S.; Wang, X.; Zhuo, H.; Gang, D.D. Analysis and Risk Assessment of Organic Pollutants in Surface Water from Xujiahe Basin, China. *Bull. Environ. Contam. Toxicol.* **2020**, *105*, 453–459. [CrossRef] [PubMed]
4. Jiang, J.; Zhao, J.; Zhao, G.; Liu, L.; Song, H.; Liao, S. Recognition, possible source, and risk assessment of organic pollutants in surface water from the Yongding River Basin by non-target and target screening. *Environ. Pollut.* **2023**, *331*, 121895. [CrossRef] [PubMed]
5. Bi, S.; Wang, L.; Li, Y.R.; Zhang, Z.P.; Wang, Z.M.; Ding, X.L.; Zhou, J.Z. A Comprehensive Method for Water Environment Assessment considering Trends of Water Quality. *Adv. Civ. Eng.* **2021**, *2021*, 5548113. [CrossRef]
6. He, X.; Li, A.; Wang, S.; Chen, H.; Yang, Z. Perfluorinated substance assessment in sediments of a large-scale reservoir in Danjiangkou, China. *Environ. Monit. Assess.* **2018**, *190*, 66. [CrossRef] [PubMed]
7. Lin, L.; Pan, X.; Zhang, S.; Li, D.; Zhai, W.; Wang, Z.; Tao, J.; Mi, C.; Li, Q.; Crittenden, J.C. Distribution and source of microplastics in China's second largest reservoir-Danjiangkou Reservoir. *J. Environ. Sci.* **2021**, *102*, 74–84. [CrossRef]
8. Sun, H.; Giesy, J.P.; Jin, X.; Wang, J. Tiered probabilistic assessment of organohalogen compounds in the Han River and Danjiangkou Reservoir, central China. *Sci. Total Environ.* **2017**, *586*, 163–173. [CrossRef]

9. Luo, Y.; Shi, W.; You, M.; Zhang, R.; Li, S.; Xu, N.; Sun, W. Polybrominated diphenyl ethers (PBDEs) in the Danjiangkou Reservoir, China: Identification of priority PBDE congeners. *Environ. Sci. Pollut. Res.* **2021**, *28*, 12587–12596. [CrossRef]
10. Jin, H.; Yu, C.; Lin, L.; Cheng, J.; Qin, H.; Tao, J.; Deng, S. Pollution levels and ecological risks of PPCPs in water and sediment samples of Danjiangkou Reservoir. *Environ. Sci. Pollut. Res.* **2024**, *31*, 30163–30173. [CrossRef]
11. Chen, Z.-l.; Dong, F.-S.; Xu, J.; Liu, X.-G.; Zheng, Y.-Q. Management of pesticide residues in China. *J. Integr. Agric.* **2015**, *14*, 2319–2327. [CrossRef]
12. Zheng, S.; Chen, B.; Qiu, X.; Chen, M.; Ma, Z.; Yu, X. Distribution and risk assessment of 82 pesticides in Jiulong River and estuary in South China. *Chemosphere* **2016**, *144*, 1177–1192. [CrossRef] [PubMed]
13. Wang, R.; Zhang, S.; Xiao, K.; Cai, M.; Liu, H. Occurrence, sources, and risk assessment of pyrethroid insecticides in surface water and tap water from Taihu Lake, China. *J. Environ. Manag.* **2023**, *325*, 116565. [CrossRef] [PubMed]
14. Zhou, Y.; Wu, J.; Wang, B.; Duan, L.; Zhang, Y.; Zhao, W.; Wang, F.; Sui, Q.; Chen, Z.; Xu, D.; et al. Occurrence, source and ecotoxicological risk assessment of pesticides in surface water of Wujin District (northwest of Taihu Lake), China. *Environ. Pollut.* **2020**, *265*, 114953. [CrossRef] [PubMed]
15. Han, J.; Liang, Y.; Zhao, B.; Wang, Y.; Xing, F.; Qin, L. Polycyclic aromatic hydrocarbon (PAHs) geographical distribution in China and their source, risk assessment analysis. *Environ. Pollut.* **2019**, *251*, 312–327. [CrossRef]
16. Han, B.; Zheng, L.; Lin, F. Risk assessment and source apportionment of PAHs in surface sediments from Caofeidian Long Island, China. *Mar. Pollut. Bull.* **2019**, *145*, 42–46. [CrossRef]
17. Chen, C.-F.; Ju, Y.-R.; Su, Y.-C.; Lim, Y.C.; Kao, C.-M.; Chen, C.-W.; Dong, C.-D. Distribution, sources, and behavior of PAHs in estuarine water systems exemplified by Salt River, Taiwan. *Mar. Pollut. Bull.* **2020**, *154*, 111029. [CrossRef]
18. Han, B.; Li, Q.; Liu, A.; Gong, J.; Zheng, L. Polycyclic aromatic hydrocarbon (PAH) distribution in surface sediments from Yazhou Bay of Sanya, South China, and their source and risk assessment. *Mar. Pollut. Bull.* **2021**, *162*, 111800. [CrossRef]
19. Han, B.; Liu, A.; He, S.; Li, Q.; Zheng, L. Composition, content, source, and risk assessment of PAHs in intertidal sediment in Shilaoren Bay, Qingdao, China. *Mar. Pollut. Bull.* **2020**, *159*, 111499. [CrossRef]
20. Tong, Y.; Chen, L.; Liu, Y.; Wang, Y.; Tian, S. Distribution, sources and ecological risk assessment of PAHs in surface seawater from coastal Bohai Bay, China. *Mar. Pollut. Bull.* **2019**, *142*, 520–524. [CrossRef]
21. Zhao, Y.; Li, J.; Qi, Y.; Guan, X.; Zhao, C.; Wang, H.; Zhu, S.; Fu, G.; Zhu, J.; He, J. Distribution, sources, and ecological risk assessment of polycyclic aromatic hydrocarbons (PAHs) in the tidal creek water of coastal tidal flats in the Yellow River Delta, China. *Mar. Pollut. Bull.* **2021**, *173*, 113110. [CrossRef] [PubMed]
22. Wu, B.; Zhang, Y.; Zhang, X.-X.; Cheng, S.-P. Health risk assessment of polycyclic aromatic hydrocarbons in the source water and drinking water of China: Quantitative analysis based on published monitoring data. *Sci. Total Environ.* **2011**, *410–411*, 112–118. [CrossRef] [PubMed]
23. Cao, F.; Li, Z.; He, Q.; Lu, S.; Qin, P.; Li, L. Occurrence, spatial distribution, source, and ecological risk assessment of organochlorine pesticides in Dongting Lake, China. *Environ. Sci. Pollut. Res.* **2021**, *28*, 30841–30857. [CrossRef] [PubMed]
24. Chen, Y.; Yu, K.; Hassan, M.; Xu, C.; Zhang, B.; Gin, K.Y.-H.; He, Y. Occurrence, distribution and risk assessment of pesticides in a river-reservoir system. *Ecotoxicol. Environ. Saf.* **2018**, *166*, 320–327. [CrossRef]
25. Chen, C.; Zou, W.; Cui, G.; Tian, J.; Wang, Y.; Ma, L. Ecological risk assessment of current-use pesticides in an aquatic system of Shanghai, China. *Chemosphere* **2020**, *257*, 127222. [CrossRef]
26. Black, G.P.; Woodward, E.E.; Sanders, C.J.; Gross, M.S.; Hladik, M.L. Multiresidue extraction of current-use pesticides from complex solid matrices using energized dispersive guided extraction with analysis by gas and liquid chromatography tandem mass spectroscopy. *Chemosphere* **2023**, *327*, 138550. [CrossRef]
27. USEPA. *Determination of Polycyclic Aromatic Hydrocarbons (PAHs) in Ambient Air Using Gas Chromatography/Mass Spectrometry (GC/MS)*; EPA Office of Water: Washington, DC, USA, 1999.
28. Kong, J.; Ma, T.; Cao, X.; Li, W.; Zhu, F.; He, H.; Sun, C.; Yang, S.; Li, S.; Xian, Q. Occurrence, partition behavior, source and ecological risk assessment of nitro-PAHs in the sediment and water of Taige Canal, China. *J. Environ. Sci.* **2023**, *124*, 782–793. [CrossRef]
29. Gao, X.; Li, J.; Wang, X.; Zhou, J.; Fan, B.; Li, W.; Liu, Z. Exposure and ecological risk of phthalate esters in the Taihu Lake basin, China. *Ecotoxicol. Environ. Saf.* **2019**, *171*, 564–570. [CrossRef]
30. Li, W.; Wang, L.; Wang, X.; Liu, R. Derivation of predicted no effect concentration and ecological risk assessment of polycyclic musks tonalide and galaxolide in sediment. *Ecotoxicol. Environ. Saf.* **2022**, *229*, 113093. [CrossRef]
31. USEPA. *Risk Assessment Guidance for Superfund Volume I: Human Health. Evaluation Manual (Part E, Supplemental Guidance for Dermal Risk Assessment)*; Environmental Protection Agency: Washington, DC, USA, 2004.
32. Shen, B.; Wu, J.; Zhan, S.; Jin, M. Residues of organochlorine pesticides (OCPs) and polycyclic aromatic hydrocarbons (PAHs) in waters of the Ili-Balkhash Basin, arid Central Asia: Concentrations and risk assessment. *Chemosphere* **2021**, *273*, 129705. [CrossRef]
33. Ma, W.; Hu, J.; Li, J.; Li, J.; Wang, P.; Okoli, C.P. Distribution, source, and health risk assessment of polycyclic aromatic hydrocarbons in the soils from a typical petroleum refinery area in south China. *Environ. Monit. Assess.* **2022**, *194*, 678. [CrossRef] [PubMed]
34. USEPA. *Guidance for Assessing Chemical Contaminant, Data for Use in Fish Advisories*, 3rd ed.; EPA Office of Water: Washington, DC, USA, 2000; Volume 1.

35. Wang, Y.; Zhang, S.; Cui, W.; Meng, X.; Tang, X. Polycyclic aromatic hydrocarbons and organochlorine pesticides in surface water from the Yongding River basin, China: Seasonal distribution, source apportionment, and potential risk assessment. *Sci. Total Environ.* **2018**, *618*, 419–429. [CrossRef] [PubMed]
36. Chen, Y.; Zhu, L.; Zhou, R. Characterization and distribution of polycyclic aromatic hydrocarbon in surface water and sediment from Qiantang River, China. *J. Hazard. Mater.* **2007**, *141*, 148–155. [CrossRef] [PubMed]
37. Sun, J.-H.; Wang, G.-L.; Chai, Y.; Zhang, G.; Li, J.; Feng, J. Distribution of polycyclic aromatic hydrocarbons (PAHs) in Henan Reach of the Yellow River, Middle China. *Ecotoxicol. Environ. Saf.* **2009**, *72*, 1614–1624. [CrossRef] [PubMed]
38. Zheng, B.; Wang, L.; Lei, K.; Nan, B. Distribution and ecological risk assessment of polycyclic aromatic hydrocarbons in water, suspended particulate matter and sediment from Daliao River estuary and the adjacent area, China. *Chemosphere* **2016**, *149*, 91–100. [CrossRef]
39. El-Maradny, A.; Radwan, I.M.; Amer, M.; Fahmy, M.A.; Mohamed, L.A.; Ibrahim, M.I.A. Spatial distribution, sources and risk assessment of polycyclic aromatic hydrocarbons in the surficial sediments of the Egyptian Mediterranean coast. *Mar. Pollut. Bull.* **2023**, *188*, 114658. [CrossRef]
40. Kurwadkar, S.; Sethi, S.S.; Mishra, P.; Ambade, B. Unregulated discharge of wastewater in the Mahanadi River Basin: Risk evaluation due to occurrence of polycyclic aromatic hydrocarbon in surface water and sediments. *Mar. Pollut. Bull.* **2022**, *179*, 113686. [CrossRef]
41. Huang, Y.; Li, K.; Liu, H.; Yuan, X.; Li, M.; Xiong, B.; Du, R.; Johnson, D.M.; Xi, Y. Distribution, sources and risk assessment of PAHs in soil from the water level fluctuation zone of Xiangxi Bay, Three Gorges Reservoir. *Environ. Geochem. Health* **2022**, *44*, 2615–2628. [CrossRef]
42. Zheng, B.; Ma, Y.; Qin, Y.; Zhang, L.; Zhao, Y.; Cao, W.; Yang, C.; Han, C. Distribution, sources, and risk assessment of polycyclic aromatic hydrocarbons (PAHs) in surface water in industrial affected areas of the Three Gorges Reservoir, China. *Environ. Sci. Pollut. Res.* **2016**, *23*, 23485–23495. [CrossRef]
43. Lin, L.; Dong, L.; Meng, X.; Li, Q.; Huang, Z.; Li, C.; Li, R.; Yang, W.; Crittenden, J. Distribution and sources of polycyclic aromatic hydrocarbons and phthalic acid esters in water and surface sediment from the Three Gorges Reservoir. *J. Environ. Sci.* **2018**, *69*, 271–280. [CrossRef]
44. Ambade, B.; Sethi, S.S.; Giri, B.; Biswas, J.K.; Bauddh, K. Characterization, Behavior, and Risk Assessment of Polycyclic Aromatic Hydrocarbons (PAHs) in the Estuary Sediments. *Bull. Environ. Contam. Toxicol.* **2022**, *108*, 243–252. [CrossRef] [PubMed]
45. Gu, W.; Xing, W.; Liang, M.; Wang, Z.; Zhang, B.; Sun, S.; Fan, D.; Wang, L. Occurrence, distribution, and risk assessment of pesticides in surface water and sediment in Jiangsu Province, China. *Environ. Sci. Pollut. Res.* **2023**, *30*, 118418–118429. [CrossRef] [PubMed]
46. Montuori, P.; Aurino, S.; Garzonio, F.; Sarnacchiaro, P.; Polichetti, S.; Nardone, A.; Triassi, M. Estimates of Tiber River organophosphate pesticide loads to the Tyrrhenian Sea and ecological risk. *Sci. Total Environ.* **2016**, *559*, 218–231. [CrossRef] [PubMed]
47. Pérez, D.J.; Iturburu, F.G.; Calderon, G.; Oyesqui, L.A.E.; De Gerónimo, E.; Aparicio, V.C. Ecological risk assessment of current-use pesticides and biocides in soils, sediments and surface water of a mixed land-use basin of the Pampas region, Argentina. *Chemosphere* **2021**, *263*, 128061. [CrossRef]
48. He, Y.; Guo, C.; Lv, J.; Deng, Y.; Xu, J. Occurrence, sources, and ecological risks of three classes of insecticides in sediments of the Liaohe River basin, China. *Environ. Sci. Pollut. Res.* **2021**, *28*, 62726–62735. [CrossRef]
49. Tang, S.-Z.; Chen, Z.-X.; Hao, Q.-R.; Hu, Y.-P.; Wang, J.-L.; Qin, D.-L.; Wang, P.; Wang, H.-T. Detection of PCBs and OCPs in the Irtysh River Water (GC-MS/MS) and ecological risk assessment. *MethodsX* **2024**, *13*, 102944. [CrossRef]
50. Chen, Y.-P.; Zhao, Y.; Zhao, M.M.; Wu, J.-H.; Wang, K.-B. Potential health risk assessment of HFRs, PCBs, and OCPs in the Yellow River basin. *Environ. Pollut.* **2021**, *275*, 116648. [CrossRef]
51. Guo, J.; Chen, W.; Wu, M.; Qu, C.; Sun, H.; Guo, J. Distribution, Sources, and Risk Assessment of Organochlorine Pesticides in Water from Beiluo River, Loess Plateau, China. *Toxics* **2023**, *11*, 496. [CrossRef]
52. Ndunda, E.N.; Madadi, V.O.; Wandiga, S.O. Organochlorine pesticide residues in sediment and water from Nairobi River, Kenya: Levels, distribution, and ecological risk assessment. *Environ. Sci. Pollut. Res.* **2018**, *25*, 34510–34518. [CrossRef]
53. Iyer, R.; Iken, B. Protein engineering of representative hydrolytic enzymes for remediation of organophosphates. *Biochem. Eng. J.* **2015**, *94*, 134–144. [CrossRef]
54. Zhang, S.W.; Wang, R.; Wang, F.; Cai, M. Assessment of currently used and restricted organophosphorus pesticides and their degradation products in urban drinking water: An investigation of eight cities in Yangtze River Delta urban agglomeration, East China. *J. Hazard. Mater. Adv.* **2023**, *9*, 100211. [CrossRef]
55. Wang, W.; Bai, J.; Zhang, G.; Wang, X.; Jia, J.; Cui, B.; Liu, X. Depth-distribution, possible sources, and toxic risk assessment of organochlorine pesticides (OCPs) in different river sediment cores affected by urbanization and reclamation in a Chinese delta. *Environ. Pollut.* **2017**, *230*, 1062–1072. [CrossRef] [PubMed]
56. Brovini, E.M.; Quadra, G.R.; Paranaíba, J.R.; Carvalho, L.; de Oliveira Pereira, R.; de Aquino, S.F. Occurrence and environmental risk assessment of 22 pesticides in Brazilian freshwaters. *Aquat. Toxicol.* **2023**, *260*, 106566. [CrossRef] [PubMed]
57. Jin, X.; Liu, Y.; Qiao, X.; Guo, R.; Liu, C.; Wang, X.; Zhao, X. Risk assessment of organochlorine pesticides in drinking water source of the Yangtze River. *Ecotoxicol. Environ. Saf.* **2019**, *182*, 109390. [CrossRef] [PubMed]
58. Feng, C.; Xia, X.; Shen, Z.; Zhou, Z. Distribution and sources of polycyclic aromatic hydrocarbons in Wuhan section of the Yangtze River, China. *Environ. Monit. Assess.* **2007**, *133*, 447–458. [CrossRef]

59. Zhang, T.; Yang, W.-L.; Chen, S.-J.; Shi, D.-L.; Zhao, H.; Ding, Y.; Huang, Y.-R.; Li, N.; Ren, Y.; Mai, B.-X. Occurrence, sources, and ecological risks of PBDEs, PCBs, OCPs, and PAHs in surface sediments of the Yangtze River Delta city cluster, China. *Environ. Monit. Assess.* **2014**, *186*, 5285–5295. [CrossRef]
60. Sun, W.; Niu, X.; Yin, X.; Duan, Z.; Xing, L.; Liu, A.; Ma, Y.; Gao, P. Historical evolution of polycyclic aromatic hydrocarbon pollution in Chaihe Reservoir from 1863 to 2018. *J. Environ. Manag.* **2023**, *328*, 116944. [CrossRef]
61. Li, W.-H.; Tian, Y.-Z.; Shi, G.-L.; Guo, C.-S.; Li, X.; Feng, Y.-C. Concentrations and sources of PAHs in surface sediments of the Fenhe reservoir and watershed, China. *Ecotoxicol. Environ. Saf.* **2012**, *75*, 198–206. [CrossRef]
62. Ambade, B.; Sethi, S.S.; Kurwadkar, S.; Mishra, P.; Tripathee, L. Accumulation of polycyclic aromatic hydrocarbons (PAHs) in surface sediment residues of Mahanadi River Estuary: Abundance, source, and risk assessment. *Mar. Pollut. Bull.* **2022**, *183*, 114073. [CrossRef]
63. Hale, S.E.; Lehmann, J.; Rutherford, D.; Zimmerman, A.R.; Bachmann, R.T.; Shitumbanuma, V.; O'Toole, A.; Sundqvist, K.L.; Arp, H.P.H.; Cornelissen, G. Quantifying the Total and Bioavailable Polycyclic Aromatic Hydrocarbons and Dioxins in Biochars. *Environ. Sci. Technol.* **2012**, *46*, 2830–2838. [CrossRef]
64. Tavakoly Sany, S.B.; Hashim, R.; Salleh, A.; Rezayi, M.; Mehdinia, A.; Safari, O. Polycyclic Aromatic Hydrocarbons in Coastal Sediment of Klang Strait, Malaysia: Distribution Pattern, Risk Assessment and Sources. *PLoS ONE* **2014**, *9*, e94907. [CrossRef] [PubMed]
65. Khuman, S.N.; Chakraborty, P.; Cincinelli, A.; Snow, D.; Kumar, B. Polycyclic aromatic hydrocarbons in surface waters and riverine sediments of the Hooghly and Brahmaputra Rivers in the Eastern and Northeastern India. *Sci. Total Environ.* **2018**, *636*, 751–760. [CrossRef] [PubMed]
66. Li, W.-G.; Huang, D.-Y.; Chen, D.; Wang, C.; Wei, G.-L. Temporal–spatial distribution of synthetic pyrethroids in overlying water and surface sediments in Guangzhou waterways: Potential input mechanisms and ecological risk to aquatic systems. *Environ. Sci. Pollut. Res.* **2019**, *26*, 17261–17276. [CrossRef] [PubMed]
67. Zhao, X.; Chen, L.; Guo, W.; Lu, S. Temporal trends, sources, and ecological risk of residual organochlorine pesticides (OCPs) in sediment core from the Dongping Lake, North China. *Environ. Sci. Pollut. Res.* **2023**, *30*, 103033–103043. [CrossRef]
68. Siddique, S.; Chaudhry, M.N.; Ahmad, S.R.; Nazir, R.; Javed, R.; Hafeez, M.R.; Alraey, Y.; Sivasamugham, L.A.; Mahmood, A. Risk surveillance with spatial distribution of Organochlorine Pesticides (OCPs) from sedimentary samples of Chenab River. *Sci. Total Environ.* **2024**, *912*, 169256. [CrossRef]
69. Li, J.; He, J.; Li, Y.; Liu, Y.; Li, W.; Wu, N.; Zhang, L.; Zhang, Y.; Niu, Z. Assessing the threats of organophosphate esters (flame retardants and plasticizers) to drinking water safety based on USEPA oral reference dose (RfD) and oral cancer slope factor (SFO). *Water Res.* **2019**, *154*, 84–93. [CrossRef]
70. Yang, Y.; Yun, X.; Liu, M.; Jiang, Y.; Li, Q.X.; Wang, J. Concentrations, distributions, sources, and risk assessment of organochlorine pesticides in surface water of the East Lake, China. *Environ. Sci. Pollut. Res.* **2014**, *21*, 3041–3050. [CrossRef]

toxics

MDPI

Article

DEHP-Induced Glioblastoma in Zebrafish Is Associated with Circadian Dysregulation of PER3

Shuhui Men [1,†], Jiayun Xu [2,†], Zhanhong Yang [3,*] and Zhenguang Yan [1,*]

1 State Key Laboratory of Environmental Criteria and Risk Assessment, Chinese Research Academy of Environmental Sciences, Beijing 100012, China; sh_men@163.com
2 Ningbo Clinical Pathology Diagnosis Center, Ningbo 315021, China; xujiayun19@mails.ucas.edu.cn
3 Environmental Standards Institute of Ministry of Ecology and Environment of the People's Republic of China, Beijing 100012, China
* Correspondence: zhanhong_yang@126.com (Z.Y.); zgyan@craes.org.cn (Z.Y.)
† These authors contributed equally to this work.

Abstract: DEHP is a plasticizer that is widely found in our water environment and poses a significant risk to the environment and human health. Long-term exposure to DEHP can cause endocrine disruption and interfere with the organism's normal functioning. In order to explore the potential effects of DEHP on the development of biological brain tissues, this study used bioinformatics analysis to confirm the diagnostic and prognostic value of PER3 in gliomas and further validated the neurotoxicity of DEHP using methods such as behavioral experiments and molecular biology in zebrafish. The experimental findings revealed that the expression level of PER3 in diseased tissues was significantly lower than that in the control group. In addition, the expression level of PER3 was significantly correlated with immune cell infiltration, immune checkpoint genes, and oncogenes. Moreover, the ROC curve analysis showed that PER3 could accurately differentiate between GBM tissues and adjacent normal tissues. To further validate the neurotoxicity of DEHP, we analyzed the effects of DEHP exposure on zebrafish development and PER3 expression by behavioral experiments and molecular biology. The results showed that exposure to DEHP substantially altered both the behavioral responses and the gene expression profiles within the brain tissues of zebrafish. PCR results indicate that the expression of circadian rhythm factor PER3 was significantly reduced in the brains of zebrafish in the exposed group, and circadian dysregulation had a certain promoting effect on the development of glioma. The aim of this work was to investigate the potential effects of DEHP contamination in a water environment on organism brain development. It was demonstrated that PER3 is an effective early diagnostic marker, which is of great significance in the diagnosis and clinical prognosis of glioma, and that DEHP exposure can lead to a significant reduction in PER3 expression in zebrafish brain tissue. This study further proved that DEHP has a potential carcinogenic effect, which adds scientific evidence to the carcinogenicity study of DEHP.

Keywords: DEHP; carcinogenic effect; circadian rhythm; Period circadian regulator 3

1. Introduction

Di-2-ethylhexyl phthalate (DEHP) is a plasticizer that is commonly used in the production of food packaging and medical supplies. Due to its unstable covalent bond, it can easily be released into the environment from plastic products, resulting in environmental estrogenic pollution. The direct sources of DEHP in the water environment are the discharge of industrial and domestic wastewater; the indirect sources are DEHP in the atmosphere, which enters the water body through wet deposition during long-range transport, and DEHP in the soil, which enters the water body through surface runoff and other pathways [1,2]. Studies have shown that long-term exposure to plasticizers can affect lipid metabolism in living organisms and cause hormone secretion disorders and neurodevelopmental abnormalities [3,4]. Neuroinflammation caused by overactivation of neuroimmune cells due to PAEs has been extensively studied in the field of toxicology [5]. Zhou et al. [6] discovered that DBP can cause learning disabilities and anxiety behaviors in mice by activating microglia in hippocampal neurons. Similarly, Win Shwe et al. [7] exposed mice to DEHP and found that it may induce neuroinflammation by modulating hypothalamic neuroimmune biomarkers. In addition, some studies have proposed a possible link between DEHP exposure and cancer development, particularly by affecting telomere length as one of the possible mechanisms [8]. Moreover, some studies have observed that chronic exposure to DEHP induces carcinogenic effects in rodents [9].

Yi Qing et al. [10] found that prenatal PAE exposure may affect circadian regulation in adult rats by altering the expression of circadian-related genes in the SCN nuclear region of the hypothalamus. Circadian rhythms are crucial for maintaining the health of biological organisms as they regulate a wide range of physiological functions, including metabolism, sleep, endocrine, immune, and cardiovascular functions [11,12]. In addition, there is a correlation between circadian rhythms and tumor development. Studies have shown that the dysregulation of CRY1 in mice has been linked to the development of gliomas, indicating that circadian dysregulation contributes to brain tissue development [13,14]. It has been shown that circadian dysregulation can accelerate tumor development through endocrine mechanisms such as altered clock gene expression and suppression of melatonin secretion [15]. Furthermore, circadian rhythms may indirectly affect psychological recovery in cancer patients through clusters of cancer-related symptoms [16].

Additionally, the Period gene family of genes encodes circadian components related to motor activity and metabolism. Specifically, dysregulation of Period circadian regulator 3 (PER3) has been associated with the development of a variety of malignancies, including prostate, colorectal, and hematopoietic malignancies [17]. It was also shown that breast cancer susceptibility was significantly increased in PER3-deficient mouse models [18]. While little research has been conducted on the role of them in brain tissue development. Currently, the International Agency for Research on Cancer (IARC) of the World Health Organization has classified DEHP as a group 2B carcinogen, and there is limited evidence for the carcinogenicity of DEHP in humans and insufficient evidence for its carcinogenicity in experimental animals. Consequently, the aim of this study is to investigate the diagnostic and prognostic role of PER3 in glioblastoma multiforme (GBM) and to provide a new scientific basis for the carcinogenicity of DEHP.

2. Materials and Methods

2.1. Bioinformatics Data Collection

The RNAseq data and corresponding clinicopathologic information of 706 GBM patients were obtained from the Glioma Project of the TCGA database (https://portal.gdc.cancer.gov/; accessed on 23 November 2023) in the FPKM (fragments per kilobase per million) format. Subsequently, the RNA sequencing data were converted to the TPM (transcripts per kilobase million) format. Furthermore, three additional GBM datasets were acquired from the GEO database (https://www.ncbi.nlm.nih.gov/geo/; accessed on 23 November 2023).

2.2. Bioinformatics Analysis of Period3 mRNA Expression Levels

The mRNA expression levels of PER3 were compared and analyzed in 33 human cancers from the TCGA database as well as in normal tissue samples. Additionally, the expression of PER3 in glioblastoma samples was analyzed using gene expression data from the GSE29458, GSE22866, and GSE14805 datasets from the GEO database. The case samples were divided into high- and low-expression groups to analyze the differentially expressed genes (DEGs) between the two groups based on the median expression value of PER3. A threshold parameter of absolute value of logarithmic fold change (Log2FC) greater than 1.5 and p-value less than 0.05 was used. Volcano and heat maps were visualized using the "ggplot2" (v3.3.6) R package.

2.3. Functional Enrichment Analysis

Gene symbols were converted to Entrez IDs using the R package "org.hs.egg.db" (v3.10.0). Functional annotation and gene set enrichment analysis (GSEA) of differential genes were performed using the R package "ClusterProfiler" (v3.14.3). The reference gene sets used in the GSEA analysis were organized from the MSigDB database (https://www.gsea-msigdb.org/gsea/msigdb/index.jsp; accessed on 29 November 2023), and c2.cp.v7.2.symbols.gmt was selected. We screened 1518 gene sets that were significantly enriched based on the threshold parameters of FDR less than 0.25 and *p* adjust less than 0.05.

2.4. Correlation Analysis of PER3 Expression Level, Immune Cell Infiltration and Immune Cell Markers

The immune cell tumor infiltration status was assessed using the ssGSEA algorithm from the "GSVA" (v1.34.0) R package, which includes 24 types of immune cells, such as neutrophils, macrophages, Th2 cells, NK CD56dim cells, TFH, Tem, Tgd, and Tcm. Spearman correlation analysis was performed to determine the relationship between PER3 expression levels, immune cell infiltration status, and immune cell markers. Furthermore, we analyzed the correlation between PER3 expression levels and immune checkpoint genes as well as oncogenes in GBM samples from the TCGA database using the R package "ggplot2" (v3.3.6). A correlation was considered significant if the *p*-value was less than 0.05.

2.5. Analysis of DNA Methylation Status and Genetic Alterations

The MethSurv database (https://biit.cs.ut.ee/methsurv/; accessed on 11 January 2024) was used to analyze the DNA methylation status of the CpG site of the PER3 gene in the GBM-TCGALGG dataset. The prognostic value of PER3's CPG methylation status was also assessed in GBM samples, as well as its relationship with overall survival (OS) in GBM.

The PER3 genome changes were analyzed using cBioPortal (https://www.cbioportal.org/; accessed on 11 January 2024) [19,20] in three datasets: Mayo Clinic, Clinical Cancer Research 2020; CPTAC, Cell 2021; and TCGA, Firehose Legacy. The prognostic significance of PER3 gene alterations was determined by K-M survival curve analysis and log-rank test. A *p*-value of less than 0.05 was used as the criterion for a significant difference.

2.6. Correlation Analysis of PER3 Expression Level and Pathologic Characteristics

The TCGA-GBMLGG dataset was used to obtain pathological data of GBM patients, which included overall survival (OS), disease-specific survival (DSS), and progression-free interval (PFI). The statistical data were compared between the high and low expression groups in various pathological parameters such as age, WHO classification, histological type, OS events, DSS events, PFI events, 1p/19q deletion or not, IDH type, and tumor efficacy assessment using the "ggplot2" (v3.3.6) R package. The Shapiro–Wilk normality test was used to analyze normally distributed data ($p < 0.05$). The study utilized logistic regression analysis to assess the correlation between PER3 expression levels and the pathological characteristics of patients with GBM.

2.7. Prognostic Implications of PER3 Expression in Brain

In this study, patient survival data from the TCGA-GBMLGG were analyzed to determine the survival outcomes of GBM patients based on the expression levels of PER3. The analysis was performed using K-M survival curve analysis and univariate and multivariate Cox regression analysis. Diagnostic ROC curves and time-dependent ROC curves were plotted using the "pROC" (v1.18.0.), "timeROC" (v0.4), and "ggplot2" (v3.3.6) R packages. Nomogram modeling analysis was also performed to evaluate the predictive value of PER3 expression levels for GBM. Kaplan–Meier survival curves were utilized to conduct prognostic analysis on subgroups of patients with glioblastoma (GBM). The results included sample size, confidence intervals, risk ratios, and *p*-values.

2.8. Chemicals and Test Solutions

Di-2-ethylhexyl phthalate (CAS: 117-81-7, purity 99.5%; Product No. D201154) was purchased from Sigma-Aldrich (Darmstadt, Germany). The stock DEHP solutions were prepared with acetone (CAS: 67-64-1, purity 99.5%; Product No. 10000418; Sinopharm Chemical Reagent Co., Ltd. Shanghai, China). The acetone content in the final exposure solution is less than 5%.

2.9. Breeding of Zebrafish and Collection of Embryos

The AB zebrafish (*Danio rerio*) were purchased from the China Zebrafish Resource Center (Wuhai, China) and reared under the following conditions: temperature of 28 ± 0.5 °C, light/dark cycle of 14 h/10 h, and feeding of fresh plumpy shrimp twice a day.

The night before fertilized eggs were collected, males and females were placed in a spawning box in a 2:1 ratio with a baffle in the middle; the baffle was removed immediately after light exposure the following day, and eggs were collected after 30 min of spawning. Embryos/larvae were raised in E3 medium [21] (294.0 mg/L $CaCl_2 \cdot 2H_2O$, 123.3 mg/L $MgSO_4 \cdot 7H_2O$, 64.7 mg/L $NaHCO_3$, 5.7 mg/L KCl) in a Petri dish and placed in an incubator at 28 °C.

All zebrafish husbandry and experimental procedures of our study were performed in accordance with the guidelines of the Animal Care and Use Committee of Chinese Research Academy of Environmental Sciences.

2.10. Exposure Experimental Protocol

Zebrafish were divided into control group, low concentration group (50 μg/L), medium concentration group (500 μg/L), and high concentration group (1500 μg/L), where 50 μg/L was a reference to the health baseline value of DEHP [22]. A 500 μg/L amount refers to the environmental background value of the contaminated areas in China [2,23,24], and in order to better observe the toxicological mechanism, a high concentration value of 1500 μg/L is set. Each group was set up with 3 parallel groups, and 20 zebrafish in each parallel group were exposed to DEHP solution for 30 days. A new exposure solution was replaced every 24 h during the experiment.

At the end of the exposure, zebrafish embryos were obtained according to the methods described in Section 2.2. The morphological observations and behavioral experiments were performed at 5 dpf and 7dpf to observe the effects of DEHP exposure on zebrafish offspring, respectively.

2.11. Behavioral Experiments on Zebrafish Larvae

To investigate the effects of DEHP exposure on the behavioral patterns of zebrafish offspring, we controlled the light intensity of the behavioral experimental apparatus and observed the responses of zebrafish larvae in a dark environment and under alternating light and dark stimuli. Ten larvae per group were selected for behavioral observations. First, 7 dpf zebrafish larvae were placed in a well plate and acclimatized to a dark environment for 2 min. Then, their movement trajectory and residence time in the dark environment were recorded for 10 min. Next, the zebrafish were acclimatized to a bright environment for 2 min, followed by alternating light and dark stimuli performed at a frequency of 5 min for a cycle of 3 times. The study counted the average movement speed of larvae during a 30 s interval and observed behavioral differences between exposed and control zebrafish.

2.12. Quantitative Real-Time PCR

Zebrafish were euthanized using MS-222 anesthetics, followed by dissection to obtain intact zebrafish brain tissue. Each group contained 3 replicates, and each sample contained 5 brain tissues. Total RNA was extracted from each sample using MolPure® Cell/Tissue

Total RNA Kit (YEASEN, 19221ES50), and the RNA samples were reverse transcribed using Hifair® III qPCR First-Strand CDNA Synthesis Ready-to-Use Pre-Mix (YEASEN, 11141ES60) (with gDNA removal) and cDNA solution was obtained. A 20 μL system for qPCR was prepared using the SYBRGreen MasterMix (YEASEN, 11184ES08) fluorescent dye assay. The reaction was carried out using a two-step PCR program, which included denaturation at 95 °C for 10 s, annealing/extension at 60 °C for 31 s, and cycling for 40 times. Prior to this, pre-denaturation at 95 °C for 2 min was performed to promote the template to unravel the secondary structure and revitalize the hot starter enzyme. Primers were designed with Primer3plus software (3.3.0) and analyzed for specificity using NCBI Blast online.

3. Results

3.1. Expression of PER3 in Multiple Cancers Including GBM

PER3 expression was analyzed in 33 cancer datasets from the TCGA database. In 23 cancers, PER3 showed significant downregulation. Furthermore, PER3 expression was significantly reduced in glioblastoma tissues from the GSE29458, GSE22866, and GSE14805 datasets compared to the corresponding controls.

The 706 GBM patients were categorized into high- and low-expression groups based on the median PER3 expression value. The threshold parameters used were an absolute value of log fold change greater than 1.5 and a *p*-value of less than 0.05. A total of 2284 genes were found to be significantly different compared to the PER3 expression value, with 2125 genes being significantly downregulated and 158 genes being significantly upregulated. The single gene co-expression heatmap in Figure 1F displays ten of the most significantly differentiated genes.

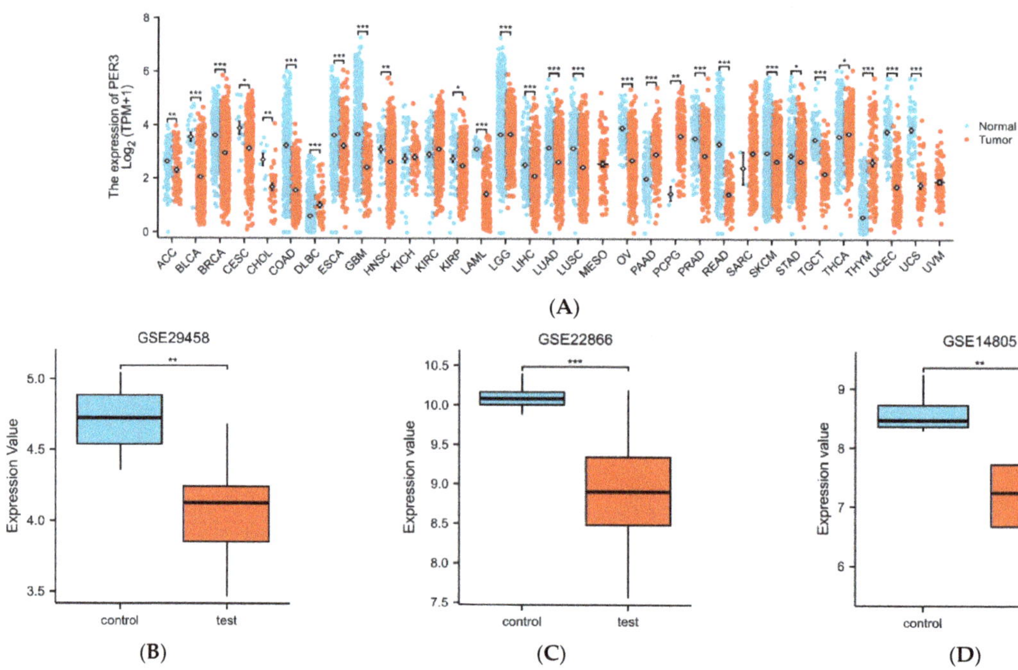

Figure 1. *Cont.*

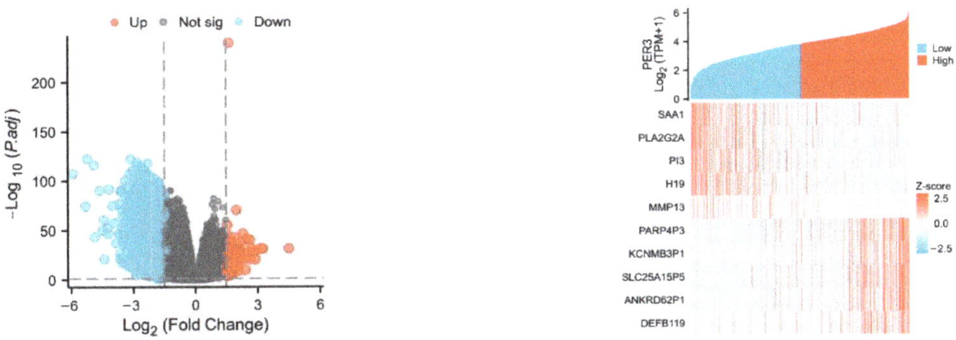

(**E**) (**F**)

Figure 1. PER3 expression is significantly downregulated in several cancers, including GBM. (**A**) TCGA database analysis showed the expression levels of PER3 in 33 cancer tissues and their corresponding surrounding normal tissues. * $p < 0.05$; ** $p < 0.01$; *** $p < 0.001$. (**B–D**) The expression levels of PER3 in cases (**B**) GSE29458, (**C**) GSE22866, and (**D**) GSE14805 in GBM tissues had significantly higher expression levels than normal brain tissues. (**E,F**) The 706 GBM patients in the TCGA-GBMLGG database were categorized into high-expression and low-expression groups based on the median PER3 expression level. (**E**) Volcano plot and (**F**) heat map demonstrating the expression levels of specific mRNAs in GBM patients.

3.2. Functional Enrichment Analysis of PER3-Related Differentially Expressed Genes in GBM

Functional annotation of differential genes associated with PER3 in GBM patients was performed using ClusterProfiler R. GO enrichment analysis results revealed highly enriched biological processes, cellular components, and molecular functions ($p < 0.05$), as shown in Figure 2A and Supplementary Materials. The biological processes that ranked the highest were phagocytosis, humoral immune response, complement activation, and immune response mediated by circulating immunoglobulin. The cellular component GO terms that were most enriched were immunoglobulin complex, blood microparticle, and the external side of the plasma membrane. The molecular functions that were most enriched were antigen binding, immunoglobulin receptor binding, receptor ligand activity, and signaling receptor activator activity.

The Gene Set Enrichment Analysis (GSEA) can reveal functional and pathway distinctions between the tumor and normal groups. Compared to general difference analysis (GO and KEGG), GSEA analysis does not require the delineation of a difference multiplicity threshold. Instead, it orders and assigns enrichment to genes as a whole based on their expression. This approach more comprehensively and intuitively demonstrates the overall expression trend of the gene set [25]. The study found that the set of PER3-related differential genes was significantly enriched for genes related to neural function, including neurexins and neuroligins (NES = 3.332, padj < 0.001, FDR < 0.001), protein interactions at synaptic level (NES = 3.213, padj < 0.001, FDR < 0.001), neuron markers (NES = 3.148, padj < 0.001, FDR < 0.001), dopamine neurotransmitter release cycle (NES = 2.937, padj < 0.001, FDR < 0.001), activation of Nmda receptors and postsynaptic events (NES = 2.613, padj < 0.001, FDR < 0.001), GABA receptor activation (NES = 2.367, padj < 0.001, FDR < 0.001), etc. In addition, PER3-associated DEGs were also enriched in neurological disease-associated gene sets, such as glioblastoma neural (NES = 3.378, padj < 0.001, FDR < 0.001), aging brain Dn (NES = 3.086, padj < 0.001, FDR < 0.001), glioblastoma mesenchymal (NES = −2.253, padj < 0.001, FDR < 0.001), melanoma metastasis UP (NES = −1.929, padj < 0.001, FDR < 0.001), GBM silenced by methylation (NES = −2.086, padj < 0.001, FDR < 0.001), synaptic signaling pathways associated with autism spectrum disorder (NES = 2.466, padj < 0.001, FDR < 0.001), etc.

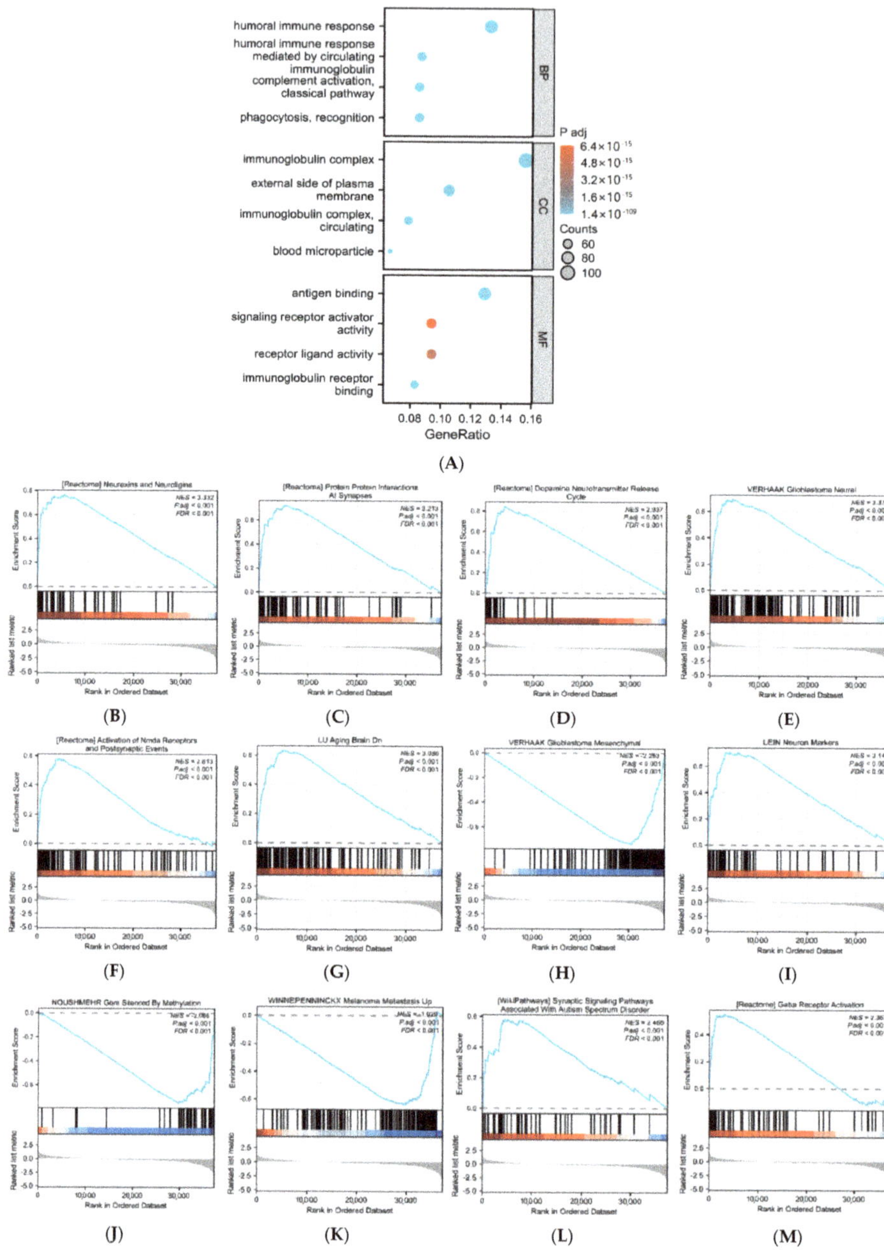

Figure 2. Functional enrichment analysis of differentially expressed genes in GBM based on PER3 expression levels. (A) GO functional enrichment analysis of PER3-associated differential genes showing enrichment for biological process (BP), cellular component (CC), and molecular function (MF). (B–M) GSEA genomic enrichment analysis of altered signaling pathways in GBM tissues based on PER3-related differential genes in GBM high and low expression groups.

Figure 2B–M displays the enrichment score in the upper part of the result graph. A higher ES value indicates that the genes are enriched in the pathway rather than dispersed.

A positive NES indicates head enrichment, with core molecules concentrated in this gene tending to be upregulated. Conversely, a negative NES indicates tail enrichment, with core molecules concentrated in this gene tending to be downregulated. The position of each gene is marked with lines in the images, while the vertical lines represent the concentration of molecules in the gene.

3.3. PER3 Expression Levels Correlate with Multiple Immune Cell Infiltration in Glioblastoma Tissues

The single-sample Gene Set Enrichment Analysis (ssGSEA) was applied to evaluate the infiltration of 24 immune cells in GBM tissues [26,27]. The relationship between PER3 and immune cell infiltration was assessed through Spearman correlation analysis. The figure shows that PER3 expression levels were negatively correlated with neutrophils (R = −0.368, $p < 0.001$), macrophages (R = −0.365, $p < 0.001$), Th2 cells (R = −0.360, $p < 0.001$), and NK CD56dim cells (R = −0.329, $p < 0.001$). Conversely, PER3 expression levels were negatively correlated with Tcm cells (R = 0.456, $p < 0.001$), Tgd cells (R = 0.367, $p < 0.001$), Tem cells (R = 0.267, $p < 0.001$), and TFH cells (R = 0.256, $p < 0.001$). We next evaluated the relationship between the infiltration levels of the above several immune cells in different PER3 expression groups, and the results obtained were consistent with the analysis in Figure 3A.

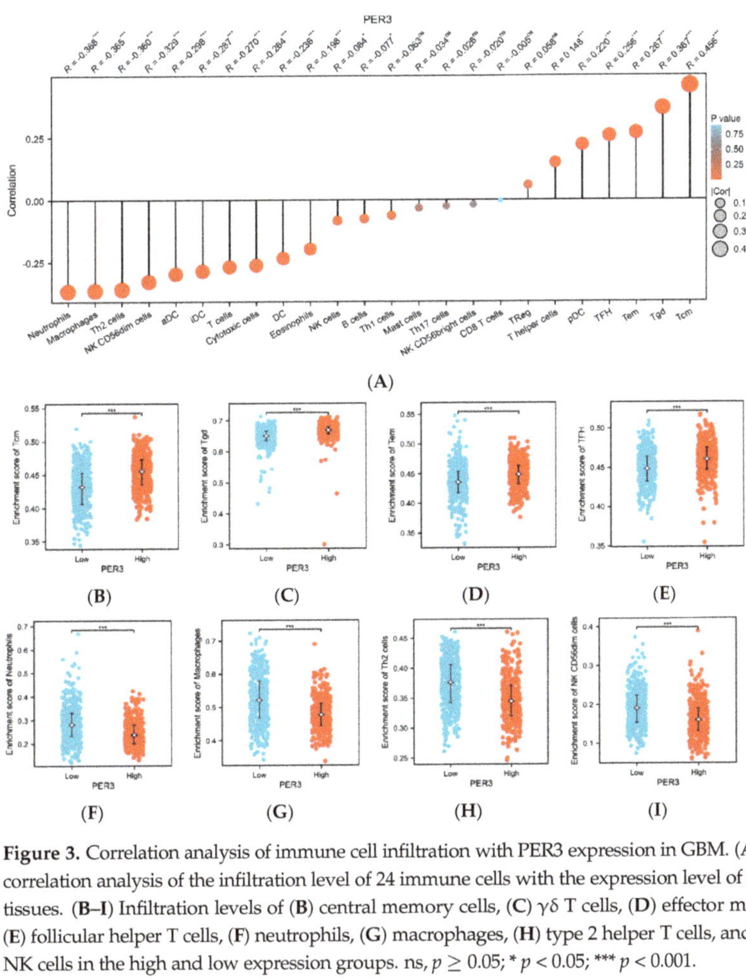

Figure 3. Correlation analysis of immune cell infiltration with PER3 expression in GBM. (**A**) Spearman's correlation analysis of the infiltration level of 24 immune cells with the expression level of PER3 in GBM tissues. (**B–I**) Infiltration levels of (**B**) central memory cells, (**C**) $\gamma\delta$ T cells, (**D**) effector memory T cells, (**E**) follicular helper T cells, (**F**) neutrophils, (**G**) macrophages, (**H**) type 2 helper T cells, and (**I**) CD56dim NK cells in the high and low expression groups. ns, $p \geq 0.05$; * $p < 0.05$; *** $p < 0.001$.

3.4. Correlation of PER3 Expression Levels with the Expression of Immune Checkpoint Genes and Oncogenes in Glioblastoma Tissue

CD276, HLA-A, and TNFRSF4 are immune checkpoint proteins that are associated with tumor immune evasion. Their receptors are co-stimulatory factors that are associated with T cell immunity. High expression of immune checkpoint molecules can suppress the body's immune function.

Figure 4 exhibits the correlation between the expression levels of PER3 and the above genes in GBM. It was found that the expression level of the PER3 gene was positively correlated with that of CPEB3 and negatively correlated with the expression of other oncogenes and immune checkpoint proteins in GBM samples in the TCGA database. These results further demonstrate the potential diagnostic role of PER3 in GBM.

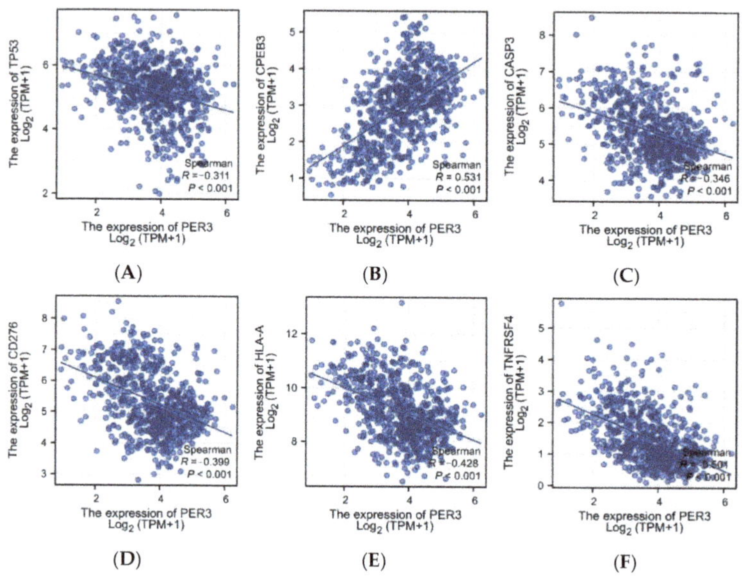

Figure 4. Correlation analysis of PER3 expression level with oncogene (**A**) TP53, (**B**) CPEB3, (**C**) CASP3, and immune checkpoint (**D**) CD276, (**E**) HLA-A, and (**F**) TNFRSF4 expression level in glioblastoma tissues.

3.5. PER3 Gene Methylation Status Is Associated with Prognosis in Patients with Glioblastoma

In the study, the DNA methylation level of the PER3 gene and the prognostic value of the CpG islands of the PER3 gene in GBM were analyzed using the MetSurv tool (https://biit.cs.ut.ee/methsurv/; accessed on 11 January 2024). As shown in Figure 5, the results revealed methylation alterations in 22 CpG islands, with cg12258811, cg14204433, cg11843502, cg06487986, cg17465881, cg11753033, and cg23927002 showing reduced DNA methylation levels. We then analyzed the prognostic impact of methylation levels at different CpG islands on GBM patients based on the KM curves of PER3 methylated sites (Table 1). The results showed that the methylation levels of eight CpG islands (cg06487986, cg08764927, cg11978441, cg12258811, cg14204433, cg17465881, cg17724687, cg23927002) were correlated with the prognosis of GBM patients ($p < 0.05$). Specifically, reduced methylation levels of PER3 at these CpG sites were associated with lower overall survival in GBM patients compared to those with higher PER3 methylation levels.

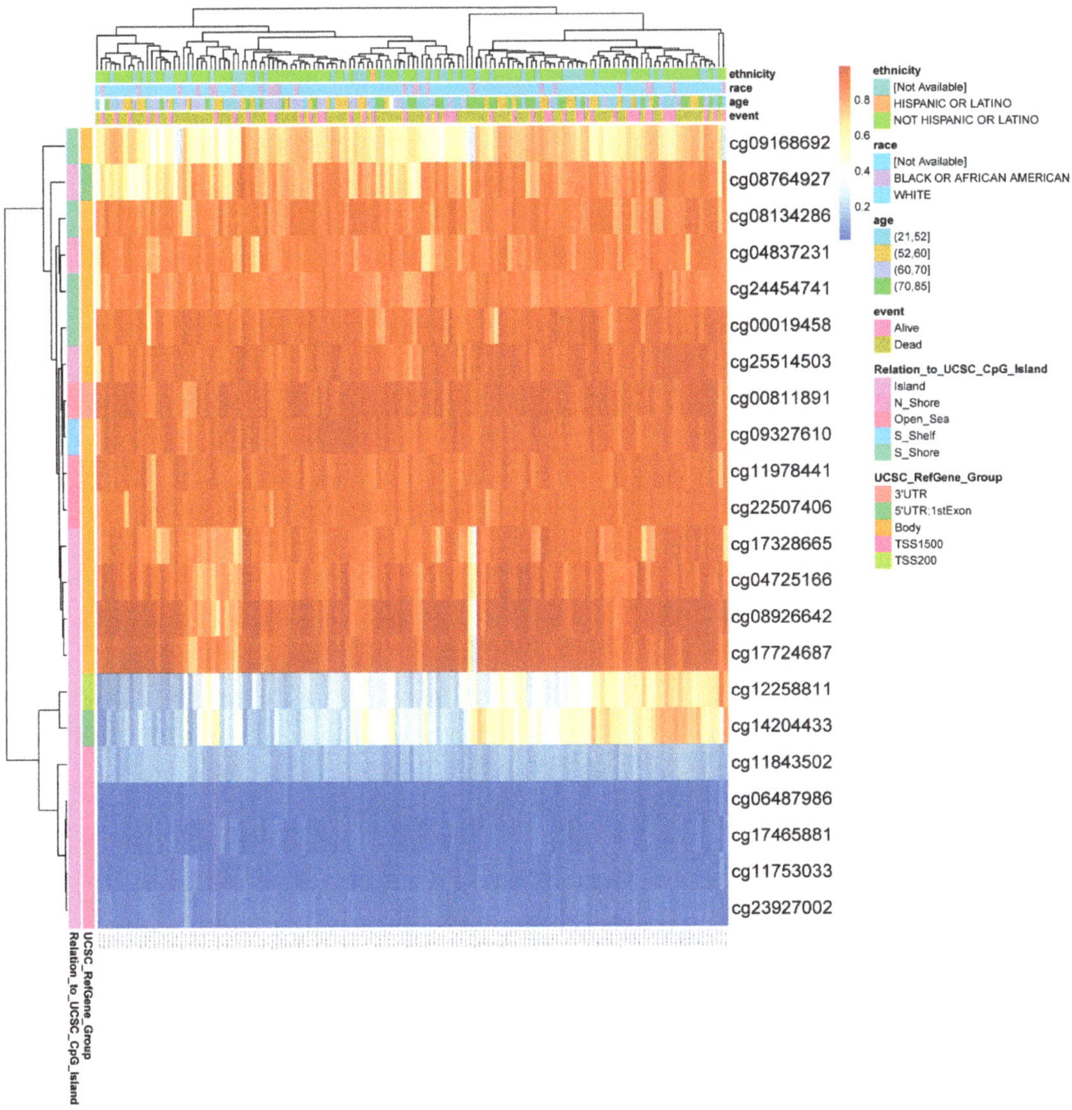

Figure 5. DNA methylation level of PER3 gene is associated with prognosis of GBM.

Table 1. Effects of methylation levels in the CpG sites of the PER3 gene on the prognosis of GBM patients.

CpG Island	HR	*p*-Value
Body-S_Shore-cg00019458	0.744	0.2004
3′UTR-Open_Sea-cg00811891	1.297	0.2570
Body-Island-cg04725166	0.701	0.1295
Body-N_Shore-cg04837231	0.683	0.1097
TSS1500-Island-cg06487986	**0.476**	**0.0011**
Body-S_Shore-cg08134286	1.160	0.5408
5′UTR;1stExon-Island-cg08764927	**0.506**	**0.0013**
Body-Island-cg08926642	0.708	0.1354

Table 1. *Cont.*

CpG Island	HR	*p*-Value
Body-S_Shore-cg09168692	0.726	0.1773
Body-S_Shelf-cg09327610	1.200	0.4652
TSS1500-Island-cg11753033	0.795	0.3440
TSS1500-Island-cg11843502	0.811	0.3937
Body-Open_Sea-cg11978441	**0.569**	**0.0156**
TSS200-Island-cg12258811	**0.524**	**0.0032**
5′UTR;1stExon-Island-cg14204433	**0.544**	**0.0174**
Body-Island-cg17328665	0.682	0.0949
TSS1500-Island-cg17465881	**0.536**	**0.0260**
Body-Island-cg17724687	**1.660**	**0.0293**
Body-Open_Sea-cg22507406	0.897	0.6443
TSS1500-Island-cg23927002	**0.602**	**0.0223**
Body-S_Shore-cg24454741	0.712	0.1076
Body-Island-cg25514503	1.321	0.2389

Bolded CpG islands are those with *p*-values less than 0.05.

3.6. PER3 Expression Levels Correlate with Multiple Clinicopathologic Features of Brain GBM

The relationship between clinical case characteristics of GBM patients and PER3 expression levels based on the TCGA-GBMLGG dataset is shown in Table 2. Significant differences were found between GBM patients with high and low PER3 expression in terms of histopathological type, WHO grade, IDH status, primary therapy outcome, OS events, DSS events, and PFI events. No significant correlation was found in terms of patients' ethnicity (Figure 6). In glioblastoma tissues, patients with high tumor progression exhibited lower levels of PER3. Logistic regression analysis revealed significant correlations between PER3 expression levels and WHO grade, IDH status, primary therapy outcome, and patient age (Table 3).

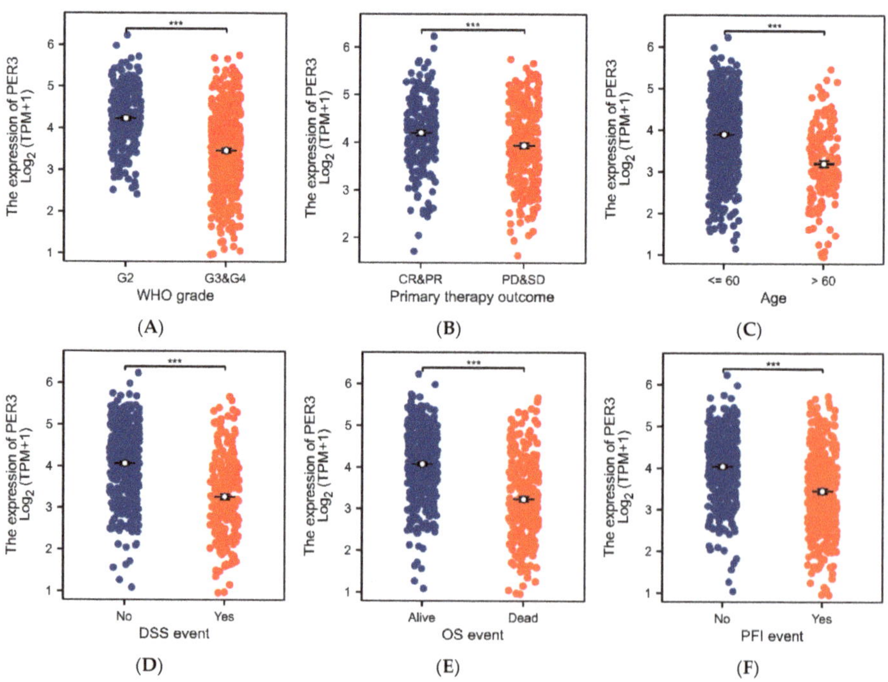

Figure 6. *Cont.*

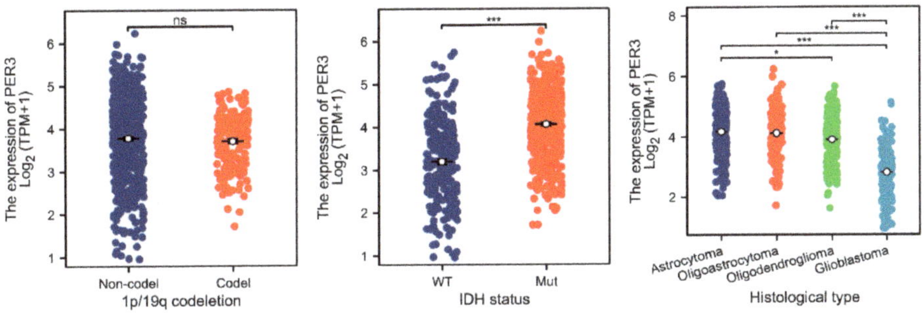

(G) (H) (I)

Figure 6. PER3 expression levels correlate with a variety of clinical case characteristics in GBM patients. (**A–I**) The correlation analysis between PER3 expression level and (**A**) WHO grade, (**B**) Primary therapy outcome, (**C**) Age, (**D**) DSS, (**E**) OS, (**F**) PEI, (**G**) 1p/19q codeletion, (**H**) IDH status, (**I**) Histological type. ns, $p \geq 0.05$; * $p < 0.05$; *** $p < 0.001$.

Table 2. Clinicopathological characteristics of GBM patients with high- and low-PER3 expression levels.

Characteristics	Low Expression of PER3	High Expression of PER3	*p*-Value
Total number of patients	349	350	
Race, *n* (%)			
Asian	6 (0.9%)	7 (1%)	0.8315
Black or African American	18 (2.6%)	15 (2.2%)	
White	318 (46.4%)	322 (46.9%)	
Histological type, *n* (%)			
Astrocytoma	66 (9.4%)	130 (18.6%)	<0.001
Glioblastoma	153 (21.9%)	15 (2.1%)	
Oligodendroglioma	85 (12.2%)	115 (16.5%)	
Oligoastrocytoma	45 (6.4%)	90 (12.9%)	
WHO grade, *n* (%)			
G2	62 (9.7%)	162 (25.4%)	<0.001
G3	109 (17.1%)	136 (21.4%)	
G4	153 (24%)	15 (2.4%)	
IDH status, *n* (%)			
WT	185 (26.9%)	61 (8.9%)	<0.001
Mut	157 (22.8%)	286 (41.5%)	
Primary therapy outcome, *n* (%)			
PD	53 (11.4%)	59 (12.7%)	0.0052
SD	62 (13.3%)	86 (18.5%)	
PR	17 (3.7%)	48 (10.3%)	
CR	42 (9%)	98 (21.1%)	
OS event, *n* (%)			
Alive	152 (21.7%)	275 (39.3%)	<0.001
Dead	197 (28.2%)	75 (10.7%)	
DSS event, *n* (%)			
No	156 (23%)	278 (41%)	<0.001
Yes	176 (26%)	68 (10%)	
PFI event, *n* (%)			
No	129 (18.5%)	224 (32%)	<0.001
Yes	220 (31.5%)	126 (18%)	

Table 3. Logistic regression analysis of the relationship between clinicopathological characteristics and the PER3 expression levels in GBM patients.

Characteristics	Total (N)	OR (95% CI)	*p* Value
WHO grade (G4 and G3 vs. G2)	637	0.221 (0.155–0.314)	<0.001
IDH status (Mut vs. WT)	689	5.525 (3.898–7.830)	<0.001
Primary therapy outcome (PD and SD vs. CR and PR)	465	0.510 (0.345–0.752)	<0.001
Gender (Male vs. Female)	699	0.915 (0.678–1.235)	0.562
Race (Black or African American and White vs. Asian)	686	0.860 (0.286–2.585)	0.788
Age (>60 vs. ≤60)	699	0.271 (0.180–0.408)	<0.001

3.7. PER3 Is a Potential Prognostic and Diagnostic Biomarker for Brain GBM

The analysis of the K-M survival curve (Figure 7) revealed that patients with glioblastoma multiforme who had low PER3 expression levels had significantly lower overall survival, disease-specific survival, and progression-free interval compared to those with high levels of PER3 expression. The results of the multifactorial Cox regression analysis, presented in Table S2, indicate that PER3 is an independent risk factor for predicting OS (HR: 0.853, $p < 0.001$). However, it is not a risk factor for DSS and PFI. The HR value less than 1 suggests that low expression of the PER3 gene is detrimental to the overall survival of GBM patients.

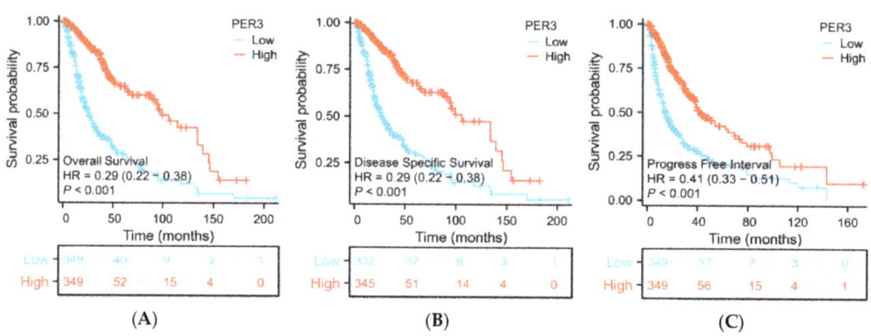

Figure 7. PER3 exhibits high prognostic predictive value in patients with GBM. k-M plots show the differences in (**A**) overall survival, (**B**) disease-specific survival, and (**C**) progression-free intervals in patients with GBM in the high and low PER3 expression groups. $p < 0.05$ was statistically significant.

Additionally, age, IDH type, and tumor efficacy assessment were independent predictors of OS, DSS, and PFI. The WHO classification was also clinically important in predicting OS and DSS.

The diagnostic value of PER3 expression level was analyzed using the receiver operating characteristic (ROC) curve (Figure 8). An area under the curve (AUC) value greater than 0.5 proved that the variable was diagnostic in predicting the outcome. The diagnostic ROC plot showed that the AUC value of PER3 as a diagnostic factor was 0.916. This indicates that the PER3 expression level can accurately differentiate the tumor tissue from the surrounding tissue adjacent to the tumor. Additionally, basic fibroblast growth factor receptor (BAFR) and epidermal growth factor receptor (EGFR), which are common molecular markers for glioblastoma, were selected as diagnostic factors for graphing. The AUC values were above 0.9, indicating that the diagnostic effect of PER3 in predicting the outcome was consistent with the common prognostic biomarkers of glioblastoma.

Time-dependent ROC curve analysis demonstrated that the AUC values for predicting 5- and 7-year survival rates of GBM patients based on PER3 expression levels were both greater than 0.7.

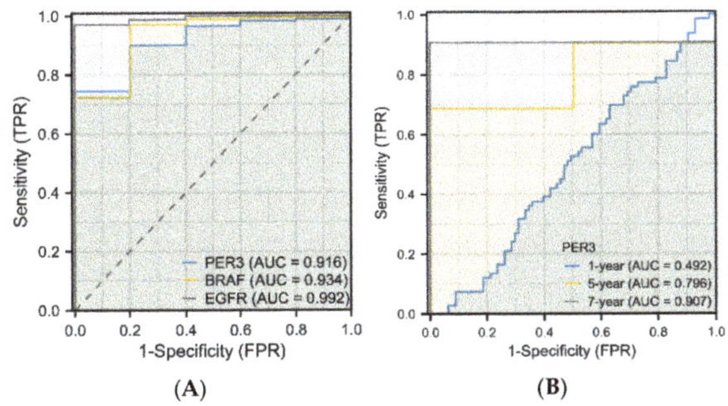

Figure 8. PER3 exhibits excellent diagnostic and prognostic performance in GBM (**A**) three-factor diagnostic ROC curve. (**B**) Time-dependent ROC curves predicting 1-, 5-, and 7-year survival rates in glioblastoma patients based on PER3 expression levels.

3.8. DEHP Exposure Effects on PER3 Gene Expression in Zebrafish Brain

The PER gene family plays a crucial role in regulating circadian rhythms. This study examined the effects of exposure to different concentrations of DEHP solution on adult zebrafish. After 30 days of exposure, the expression of the circadian factor PER3 was significantly downregulated in the zebrafish brain (Figure 9A), confirming the regulation of circadian rhythm-related genes in zebrafish by DEHP. Protein homology analysis revealed homologous similarity between zebrafish and human PER3 proteins (Figure 9E). Additionally, zebrafish are genetically similar to humans, with a genetic conservatism of 85%. This high degree of genetic similarity makes zebrafish valuable as a replacement for humans in disease or drug testing. Thus, the differential expression of the PER3 gene in zebrafish provides valuable information for studying human diseases.

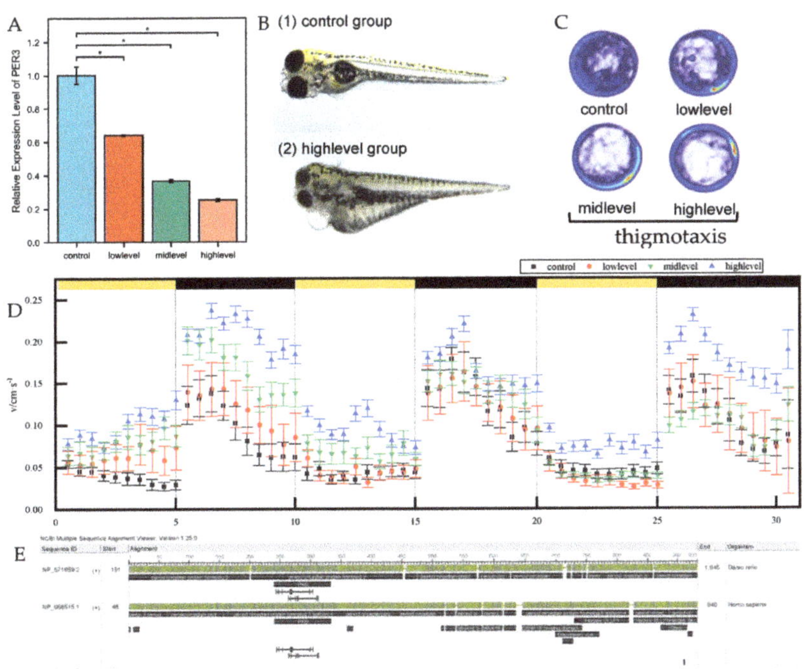

Figure 9. Toxic effects of DEHP on zebrafish. (**A**) Effects of DEHP exposure on the differential expression of PER3 gene in the zebrafish brain. * $p < 0.05$. (**B**) Effects of DEHP exposure on zebrafish

growth and development. (**C**) Effects of zebrafish exposure on spontaneous locomotion of juvenile fish. (**D**) Behavioral effects of light and dark cyclic stimulation on different subgroups of zebrafish. (**E**) Analysis of the homology of zebrafish and human PER3 protein.

3.9. Effects of DEHP Exposure on the Growth, Development, and Behavioral Patterns of Larval Zebrafish

The study observed the development of zebrafish offsprings over a 5-day period and found that DEHP exposure had a significant impact on their overall morphological development. This was mainly characterized by developmental delays, curvature of the body axis, and pericardial edema (Figure 1B).

To investigate the impact of DEHP on zebrafish circadian rhythms, we analyzed the behavior of 7 dpf zebrafish larvae. We observed their spontaneous locomotor activity in the dark and their behavior under alternating light and dark conditions. We also measured their tendency to reflect their anxiety levels in unfamiliar environments (Figure 1C). Compared to the control group, the exposed group showed a significant increase in residence time in both the peripheral and central zones. This suggests that exposure to DEHP worsened anxiety levels in zebrafish larvae.

Furthermore, the behavioral variations of zebrafish in diverse lighting conditions can serve as a foundation for identifying neurological disorders. Additionally, the response of zebrafish to changes in light and darkness can demonstrate its capacity to regulate circadian rhythms [28]. The findings indicate that exposure to DEHP can impact the reactivity of zebrafish larvae when exposed to alternating light and dark stimuli (Figure 1D). With an increase in the number of cycles, the swimming speed of the control group tends to stabilize under different conditions. However, the speed of the exposed group of juvenile fish varies more and is concentration-dependent.

4. Discussion

PER3 is a circadian regulator that can influence cell cycle, growth, and differentiation. The study found that PER3 was overexpressed in 22 out of 33 common human cancers. Researchers [18] conducted cell function experiments on the PER3 gene and found that silencing it increased the ability of breast cancer cells to proliferate, invade, and metastasize in vitro. Conversely, overexpression of PER3 had an inhibitory effect on these malignant phenotypes of breast cancer cells. Wang et al. [29] demonstrated that the expression level of PER3 was lower in both colon and rectal tumor tissues compared to normal tissues. The GEO database screening of glioblastoma cases also revealed a low expression of the PER3 gene.

The functional annotation of the differential genes showed that the differential genes related to PER3 were enriched in GBM tissues for biological processes such as humoral immune response, phagocytosis, and complement activation, cellular components such as immunoglobulin complexes, blood microparticles, and outer plasma membrane, as well as for molecular functions such as antigen binding, immunoglobulin receptor binding, receptor ligand activity, and signaling receptor activation. The GSEA results showed that genes related to PER3 were enriched in sets associated with neurological function and diseases. These findings suggest that decreased PER3 expression may disrupt normal neurological function and contribute to glioblastoma development by affecting the cycle, migration, and invasion of glioblastoma cells.

Our study demonstrated a potential relationship between aberrant expression of PER3 and tumor immune cell infiltration. The level of PER3 expression was negatively correlated with neutrophils, macrophages, Th2 cells, and NK CD56dim cells. Neutrophils play an important role in suppressing tumors by activating immune responses against tumor cells and directly lysing them [30]. Macrophages use phagocytosis as a major mechanism of innate immunity. This process enables them to engulf and kill intracellular tumor cells, as well as their own senescent and abnormal cells [31]. Macrophages are antigen-presenting cells that play a crucial role in the body's immune defense, self-stabilization, and immune surveillance by initiating the immune response. Th2 cells, a type of CD4+ T cells, contribute to humoral immunity and selectively activate macrophages, mediate the regression of the inflammatory phase, and initiate tissue repair. NK cells can inhibit tumor growth by interacting with dendritic cells, macrophages, and T cells to develop an immune response. CD56dim NK cells, which are cytotoxic, account for more than 90% of all NK cells. Our data show a positive correlation between PER3 expression and central memory T cells, $\gamma\delta$ T cells, effector memory T cells, and follicular helper T cells. In 2005, Klebanoff CA et al. [32] demonstrated the super anti-tumor ability of Tcm cells, which have the ability of self-renewal and replication and long survival in vivo. Tem cells, which are mainly located in peripheral tissues, can rapidly produce effector cytokines once re-stimulated by antigen and are considered activated Tcm cells. Tem are activated Tcm cells, mainly located in peripheral tissues, which can rapidly produce effector cytokines once they are re-stimulated by antigen. $\gamma\delta$ T cells can directly serve anti-tumor purposes and are also capable of targeting tumor cells through antibody-dependent cytotoxicity (ADCC). T follicular helper cells regulate tumor growth and progression through the chemokine receptor CXCR5 [33,34]. Taken together, our findings indicate that low expression of the PER3 gene is a significant factor in the immune evasion mechanism of GBM cells.

Abnormal expression and function of immune checkpoint molecules can contribute to the development of various diseases. For instance, overexpression or overfunction of immune checkpoint molecules can suppress immune function, weaken the body's immunity, and increase susceptibility to tumors and other diseases. Key proteins associated with immune escape from tumors include HLA-A, CD276, and TNFRSF4 [35]. CD276 is associated with diseases such as immunodeficiency and neuroblastoma. It may regulate T-cell-mediated immune responses and play a protective role in tumor cells by inhibiting natural killer-mediated cell lysis. Additionally, it can serve as a marker for the detection of neuroblastoma cells [36]. Researchers have extensively studied the pharmacological efficacy of TNFRSF4 inhibitors using a humanized mouse model. Antibodies against human OX40 have shown significant anti-tumor effects ($p < 0.001$). Furthermore, oncogenes are linked to cell proliferation and tumorigenesis. Malignant tumors exhibit high expression of TP53, and TP53 mutations are associated with poor prognosis in several human cancers [37,38]. CASP3 is a crucial indicator for detecting apoptosis, and it is associated with the development of cancers and neurological disorders [39,40]. The study showed that CPEB3 plays a crucial role in memory storage, and it affects neuronal function when its expression is inhibited [41], suggesting that the CPEB3 protein could be a promising target for treating neurodegenerative diseases. Recent studies have considered CPEB3 and CASP3 as candidate oncogenes [42].

DNA methylation is a prevalent epigenetic mechanism that plays a crucial role in tumorigenesis. Alterations in gene methylation status are linked to the development, growth, and progression of cancer. This paper examines the correlation between the methylation level of the PER3 gene and the prognosis of GBM patients. The study found that poorer overall survival was significantly associated with hypomethylation at five CpG islands, including cg23927002, cg17465881, and cg06487986, with these three loci exhibiting the lowest degree of methylation. Furthermore, the study showed that the probability of PER3 gene mutations occurring in GBM patients is less than 3%. Additionally, the study found no association between PER3 gene mutation and OS or DSS in GBM patients. Our study found significant differences in the expression level of PER3 in GBM tissues based on tumor histopathological staging, WHO grade, IDH type, tumor efficacy assessment, OS events, DSS events, and PFI events. Lower PER3 expression was correlated with poorer prognosis and later clinical staging in GBM patients.

The clinical diagnosis of glioblastoma primarily depends on imaging measures, such as CT and MRI. In early diagnosis, diagnostic ROC curves are often used to determine the diagnostic value of a particular factor for the disease and to assess the predictive accuracy, which reflects the relationship between sensitivity and specificity. This study analyzed the diagnostic value of PER3 in GBM by selecting two molecular pathology indicators of GBM from the Glioma Diagnostic and Treatment Guidelines and analyzing the ROC curves together with PER3 as a diagnostic factor. BRAF has been reported to occur in various gliomas [43], and EGFR is one of the diagnostic indicators of grade 4 GBM of the WHO [44]. Both are potential targets for targeted therapy of GBM. Both PER3 and EGFR are potential targets for glioblastoma-targeted therapy. The area under the curve (AUC) is usually taken between 0.5 and 1, with a value closer to 1 indicating greater diagnostic predictiveness. Figure 9A shows that the AUC values of PER3, BEAF, and EGFR are all above 0.9. Additionally, the AUC values of PER3 for predicting the survival rate of GBM patients at both 5 and 7 years are greater than 0.7. These data further suggest that PER3 is a potential diagnostic and prognostic biomarker for GBM [45].

Furthermore, the results of DEHP exposure experiments on zebrafish showed that long-term exposure to DEHP can lead to the adverse effects of downregulation of PER3 gene expression and circadian dysregulation in zebrafish brain tissues, suggesting that DEHP is potentially carcinogenic to organisms. Previous studies showed that the concentration of DEHP in some sections of the Yellow River was as high as 109 μg/L [23], and the highest detected concentration of DEHP in the middle and lower reaches of the Songhua River was 1752 μg/L, with an average detected concentration of 370 μg/L [24]. To summarize, the concentration of DEHP detected in the water environment in some regions of China far exceeded the limit of the standard for surface water in China (8 μg/L) [46]. Therefore, exposure of aquatic organisms in rivers to these concentrations of DEHP may cause some negative effects. The results of the study provide some theoretical basis for future management of DEHP in the water environment.

5. Conclusions

The diagnostic and prognostic value of the circadian factor PER3 in GBM was demonstrated in this study. The PER3 gene is expressed at low levels in tumor tissues of GBM patients and regulates GBM progression by modulating the expression of genes involved in the cell cycle and immune response. Additionally, the PER3 gene is associated with the tumor infiltration status of multiple immune cell types and may play a role in the response of GBM patients to immunotherapy. Taken together, these findings suggest that PER3 may serve as a potential therapeutic target and useful diagnostic and prognostic marker for GBM. Additionally, these findings provide further evidence for the carcinogenicity of DEHP. However, further in vivo and in vitro experiments are necessary to validate the results of this study.

Supplementary Materials: The following supporting information can be downloaded at: https://www.mdpi.com/article/10.3390/toxics12120835/s1, Table S1: Functional annotations of differentially expressed genes in GBM patients with different PER3 levels; Table S2: Cox regression analysis of the clinical outcomes in GBM patients based on various clinicopathological characteristics including PER3 levels.

Author Contributions: Conceptualization, S.M. and J.X.; methodology, S.M. and J.X.; validation, S.M. and J.X.; formal analysis, S.M.; investigation, S.M.; resources, S.M.; data curation, J.X. and Z.Y. (Zhenguang Yan); writing—original draft preparation, S.M.; writing—review and editing, Z.Y. (Zhenguang Yan) and Z.Y. (Zhanhong Yang); visualization, S.M. and J.X.; supervision, Z.Y. (Zhenguang Yan); funding acquisition, Z.Y. (Zhanhong Yang). All authors have read and agreed to the published version of the manuscript.

Funding: This study was financially supported by the National Key Research and Development Program of China (Grant No. 2021YFC3101703).

Institutional Review Board Statement: Not applicable.

Informed Consent Statement: Not applicable.

Data Availability Statement: The raw data supporting the conclusions of this article will be made available by the authors upon request.

Conflicts of Interest: The authors declare that they have no known competing financial interests or personal relationships that could have appeared to influence the work reported in this paper.

References

1. Wang, J.; Chen, G.; Christie, P. Occurrence and risk assessment of phthalate esters (PAEs) in vegetables and soils of suburban plastic film greenhouses. *Sci. Total Environ.* **2015**, *523*, 129–137. [CrossRef] [PubMed]
2. He, W.; Qin, N.; Kong, X. Spatio-temporal distributions and the ecological and health risks of phthalate esters (PAEs) in the surface water of a large, shallow Chinese lake. *Sci. Total Environ.* **2013**, *461*, 672–680. [CrossRef] [PubMed]
3. Swan, S.H. Environmental phthalate exposure in relation to reproductive outcomes and other health endpoints in humans. *Environ. Res.* **2008**, *108*, 177–184. [CrossRef] [PubMed]
4. Wang, Z.; Li, Z.; Lou, Q.; Pan, J.; Wang, J.; Men, S.; Yan, Z. Ecological risk assessment of 50 emerging contaminants in surface water of the Greater Bay Area. *Sci. Total Environ.* **2023**, *907*, 168105. [CrossRef]
5. Li, M.R.; Men, S.H.; Wang, Z.Y.; Liu, C.; Zhou, G.R.; Yan, Z.G. The application of human-derived cell lines in neurotoxicity studies of environmental pollutants. *Sci. Total Environ.* **2023**, *912*, 168839. [CrossRef]
6. Zhou, T.; He, S.; Ye, X.; Wei, Z.; Wan, J.; Zhang, H.; Ding, S. Exposure to dibutyl phthalate adsorbed to multi-walled carbon nanotubes causes neurotoxicity in mice by inducing the release of BDNF. *Sci. Total Environ.* **2022**, *852*, 158319. [CrossRef]
7. Win-Shwe, T.T.; Yanagisawa, R.; Koike, E.; Nitta, H.; Takano, H. Expression levels of neuroimmune biomarkers in hypothalamus of allergic mice after phthalate exposure. *J. Appl. Toxicol.* **2013**, *33*, 1070–1078. [CrossRef]
8. Cai, Q. Telomere length: A possible link between phthalate exposure and cancer development? *EBioMedicine* **2016**, *6*, 6–7. [CrossRef]
9. Agarwal, D.K.; Lawrence, W.H.; Autian, J. Antifertility and mutagenic effects in mice from parenteral administration of di-2-ethylhexyl phthalate (DEHP). *J. Toxicol. Environ. Health* **1985**, *16*, 71–84. [CrossRef]
10. Yi, Q.; Guo, N.; Zhang, L.Y.; Xu, X.X.; Sun, Z.R. Effects of prenatal exposure to Bis (2-ethylhexyl) phthalate on expression of genes associated with circadian rhythm in hypothalamus of adult offspring rats. *J. Environ. Health* **2018**, *10*, 867–870.
11. Takahashi, J.S.; Hong, H.K.; Ko, C.H.; McDearmon, E.L. The Genetics of Mammalian Circadian Order and Disorder: Implications for Physiology and Disease. *Nat. Rev. Genet.* **2008**, *9*, 764–775. [CrossRef] [PubMed]
12. Haus, E.L.; Smolensky, M.H. Shift work and cancer risk: Potential mechanistic roles of circadian disruption, light at night, and sleep deprivation. *Sleep Med. Rev.* **2013**, *17*, 273–284. [CrossRef] [PubMed]
13. Zhou, B.; Zhang, Y.; Zhang, F.; Xia, Y.; Liu, J.; Huang, R.; Wang, Y.; Hu, Y.; Wu, J.; Dai, C.; et al. CLOCK/BMAL1 regulates circadian change of mouse hepatic insulin sens itivity by SIRT1. *Hepatology* **2014**, *59*, 2196–2206. [CrossRef] [PubMed]
14. Ode, K.L.; Ukai, H.; Susaki, E.A.; Narumi, R.; Matsumoto, K.; Hara, J.; Koide, N.; Abe, T.; Kanemaki, M.T.; Kiyonari, H.; et al. Knockout-Rescue Embryonic Stem Cell-Derived Mouse Reveals Circadian-Period Control by Quality and Quantity of CRY1. *Mol. Cell* **2017**, *65*, 176–190. [CrossRef] [PubMed]
15. Li, J.Y.; Zhang, Y.M.; Wang, C.X.; Li, J.; Pan, Y. Relationship between circadian rhythm sleep-wake disorders and the development of malignant tumor. *Chin. J. Clin. Pharmacol. Ther.* **2021**, *26*, 76–81.
16. Luo, X.; Xiao, P.; Li, L.; Duan, Y.; Sk, C.A.; Xie, J. Research progress in circadian rhythms in the application of psychological rehabilitation of cancer patients. *J. Cent. South Univ.* **2022**, *47*, 1740–1747.
17. Liu, Y.F.; Yuan, G.L.; Ma, L. Progress of the PER gene family in malignant tumours. *Chin. J. Breast Dis.* **2024**, *18*, 40–46.
18. Liu, Y.; Wu, Z.; Li, Y.; Zhang, J.; Gao, Y.; Yuan, G.; Han, M. PER3 plays anticancer roles in the oncogenesis and progression of breast cancer via regulating MEK/ERK signaling pathway. *J. Chin. Med. Assoc.* **2022**, *85*, 1051–1060. [CrossRef]
19. Cerami, E.; Gao, J.; Dogrusoz, U.; Gross, B.E.; Sumer, S.O.; Aksoy, B.A.; Jacobsen, A.; Byrne, C.J.; Heuer, M.L.; Larsson, E.; et al. The cBio cancer genomics portal: An open platform for exploring multidimensional cancer genomics data. *Cancer Discov.* **2012**, *2*, 401–404. [CrossRef]
20. Gao, J.; Aksoy, B.A.; Dogrusoz, U.; Dresdner, G.; Gross, B.; Sumer, S.O.; Sun, Y.; Jacobsen, A.; Sinha, R.; Larsson, E.; et al. Integrative analysis of complex cancer genomics and clinical profiles using the cBioPortal. *Sci. Signal.* **2013**, *6*, pl1. [CrossRef]
21. E3 medium (for zebrafish embryos). *Cold Spring Harb. Protoc.* **2011**, *2011*, pdb.rec66449. [CrossRef]
22. USEPA. *Update of Human Health Ambient Water Quality Criteria: Di-2-Ethylhexyl-Phthalate*; USEPA: Washington, DC, USA, 2015.
23. Niu, J.P.; Liu, Y.P.; Ruan, Y. Pollution levels of environmental hormones in Lanzhou section of Yellow River. *J. Environ. Health* **2006**, *23*, 527–529.
24. Lu, J.L.; Hao, J.B.; Wang, C.Z. Distribution characteristics of phthalates in water in the middle and lower reaches of the Second Songhua River. *Environ. Sci. Technol.* **2007**, *30*, 35–38.
25. Subramanian, A.; Tamayo, P.; Mootha, V.K.; Mukherjee, S.; Ebert, B.L.; Gillette, M.A.; Paulovich, A.; Pomeroy, S.L.; Golub, T.R.; Lander, E.S.; et al. Gene set enrichment analysis: A knowledge-based approach for interpreting genome-wide expression profiles. *PNAS* **2005**, *102*, 15545–15550. [CrossRef]
26. Hänzelmann, S.; Castelo, R.; Guinney, J. GSVA: Gene set variation analysis for microarray and RNA-seq data. *BMC Bioinform.* **2013**, *14*, 7. [CrossRef]
27. Bindea, G.; Mlecnik, B.; Tosolini, M.; Kirilovsky, A.; Waldner, M.; Obenauf, A.C.; Angell, H.; Fredriksen, T.; Lafontaine, L.; Berger, A.; et al. Spatiotemporal dynamics of intratumoral immune cells reveal the immune landscape in human cancer. *Immunity* **2013**, *39*, 782–795. [CrossRef]
28. Liao, G.Z. Effect and mechanism of combined exposure of lead and cadmium on circadian rhythm of exercise behavior of juvenile zebrafish. *South. Univ. Sci. Technol.* **2019**, 77.

29. Wang, X.; Yan, D.; Teng, M.; Fan, J.; Zhou, C.; Li, D.; Qiu, G.; Sun, X.; Li, T.; Xing, T.; et al. Reduced Expression of PER3 Is Associated with Incidence and Development of Colon Cancer. *Ann. Surg. Oncol.* **2012**, *19*, 3081–3088. [CrossRef]
30. Wang, G.; Wang, J.; Niu, C.; Zhao, Y.; Wu, P. Neutrophils: New Critical Regulators of Glioma. *Front. Immunol.* **2022**, *4*, 927233. [CrossRef]
31. Lawrence, T.; Natoli, G. Transcriptional regulation of macrophage polarization: Enabling diversity with identity. *Nat. Rev. Immunol.* **2023**, *11*, 750–761. [CrossRef]
32. Klebanoff, C.A.; Gattinoni, L.; Restifo, N.P. Sorting through subsets: Which T-cell populations mediate highly effective adoptive immunotherapy? *J. Immunother.* **2012**, *35*, 651–660. [CrossRef] [PubMed]
33. Krenács, L.; Krenács, D.; Borbényi, Z.; Tóth, E.; Nagy, A.; Piukovics, K.; Bagdi, E. Comparison of Follicular Helper T-Cell Markers with the Expression of the Follicular Homing Marker CXCR5 in Peripheral T-Cell Lymphomas-A Reappraisal of Follicular Helper T-Cell Lymphomas. *Int. J. Mol. Sci.* **2023**, *25*, 428. [CrossRef] [PubMed]
34. Fonseca, V.R.; Graca, L. Contribution of FoxP3$^+$ Tfr cells to overall human blood CXCR5$^+$ T cells. *Clin. Exp. Immunol.* **2019**, *195*, 302–304. [CrossRef] [PubMed]
35. Hu, F.F.; Liu, C.J.; Liu, L.L.; Zhang, Q.; Guo, A.Y. Expression profile of immune checkpoint genes and their roles in predicting immunotherapy response. *Brief. Bioinform.* **2021**, *22*, 176. [CrossRef]
36. Picarda, E.; Ohaegbulam, K.C.; Zang, X. Molecular Pathways: Targeting B7-H3 (CD276) for Human Cancer Immunotherapy. *Clin. Cancer Res.* **2016**, *22*, 3425–3431. [CrossRef]
37. Jiang, Z.; Liu, Z.; Li, M.; Chen, C.; Wang, X. Immunogenomics Analysis Reveals that TP53 Mutations Inhibit Tumor Immunity in Gastric Cancer. *Transl. Oncol.* **2018**, *11*, 1171–1187. [CrossRef]
38. Xiao, W.; Du, N.; Huang, T.; Guo, J.; Mo, X.; Yuan, T.; Chen, Y.; Ye, T.; Xu, C.; Wang, W.; et al. TP53 Mutation asPoten-tial Negative Predictor for Response of Anti-CTLA-4 Therapy in Metastatic Melanoma. *EBioMedicine* **2018**, *32*, 119–124. [CrossRef]
39. Zhou, Z.; Xu, S.; Jiang, L.; Tan, Z.; Wang, J. A Systematic Pan-Cancer Analysis of CASP3 as a Potential Target for Immunotherapy. *Front. Mol. Biosci.* **2022**, *9*, 776808. [CrossRef]
40. Eskandari, E.; Negri, G.L.; Tan, S.; MacAldaz, M.E.; Ding, S.; Long, J.; Nielsen, K.; Spencer, S.E.; Morin, G.B.; Eaves, C.J. Dependence of human cell survival and proliferation on the CASP3 prodomain. *Cell Death Discov.* **2024**, *10*, 63. [CrossRef]
41. Gu, X.; Schafer, N.P.; Wang, Q.; Song, S.S.; Chen, M.; Waxham, M.N.; Wolynes, P.G. Exploring the F-actin/CPEB3 interaction and its possible role in the molecular mechanism of long-term memory. *PNAS* **2020**, *117*, 22128–22134. [CrossRef]
42. Cheng, J.; Demeulemeester, J.; Wedge, D.C.; Vollan, H.K.M.; Pitt, J.J.; Russnes, H.G.; Pandey, B.P.; Nilsen, G.; Nord, S.; Bignell, G.R.; et al. Pan-cancer analysis of ho-mozygous deletions in primary tumours uncovers rare tumour suppressors. *Nat. Commun.* **2017**, *8*, 1221. [CrossRef] [PubMed]
43. Cai, X.; Chen, Z.; Chang, B.; Tu, M.; Li, S.; Wang, X.; Chen, M. Prediction of BRAF mutation status in glioblastoma multiforme by preoperative ring enhancement appearances on MRI. *Front. Oncol.* **2022**, *12*, 937345. [CrossRef] [PubMed]
44. Higa, N.; Akahane, T.; Hamada, T.; Yonezawa, H.; Uchida, H.; Makino, R.; Watanabe, S.; Takajo, T.; Yokoyama, S.; Kirishima, M.; et al. Distribution and favorable prognostic implication of genomic EGFR alterations in IDH-wildtype glioblastoma. *Cancer Med.* **2023**, *12*, 49–60. [CrossRef] [PubMed]
45. Liu, J.; Lichtenberg, T.; Hoadley, K.A.; Poisson, L.M.; Lazar, A.J.; Cherniack, A.D.; Kovatich, A.J.; Benz, C.C.; Levine, D.A.; Lee, A.V.; et al. An Integrated TCGA Pan-Cancer Clinical Data Resource to Drive High-Quality Survival Outcome Analytics. *Cell* **2018**, *173*, 400–416. [CrossRef]
46. *GB 3838-2002*; Environmental Quality Standards for Surface Water. Ministry of Ecology and Environment of the People's Republic of China; China Environment Publishing Group: Beijing, China, 2002.

toxics

MDPI

Review

Progress and Prospects of Microplastic Biodegradation Processes and Mechanisms: A Bibliometric Analysis

Yingnan Cao [1], Jing Bian [1], Yunping Han [2,3,*], Jianguo Liu [1,*], Yuping Ma [1], Weiying Feng [4], Yuxin Deng [4] and Yaojiang Yu [1]

[1] Key Laboratory of Environmental Pollution Control and Remediation at Universities of Inner Mongolia Autonomous Region, College of Resources and Environmental Engineering, Inner Mongolia University of Technology, Hohhot 010051, China; caoyingnan@imut.edu.cn (Y.C.); 17304630142@163.com (J.B.); waipomyp99@163.com (Y.M.); 18347429190@163.com (Y.Y.)

[2] State Key Laboratory of Environmental Aquatic Chemistry, Research Center for Eco-Environmental Sciences, Chinese Academy of Sciences, Beijing 100085, China

[3] University of Chinese Academy of Sciences, Beijing 100049, China

[4] School of Materials Science and Engineering, Beihang University, Beijing 100191, China; fengweiying@buaa.edu.cn (W.F.); dengyx@buaa.edu.cn (Y.D.)

* Correspondence: yphan@rcees.ac.cn (Y.H.); jgliu_yxc@163.com (J.L.); Tel.: +86-10-62911425 (Y.H.)

Abstract: In order to visualize the content and development patterns of microplastic biodegradation research, the American Chemical Society (ACS), Elsevier, Springer Link, and American Society for Microbiology (ASM) were searched for the years 2012–2022 using Citespace and VOSvivewer for bibliometrics and visual analysis. The biodegradation processes and mechanisms of microplastics were reviewed on this basis. The results showed a sharp increase in the number of publications between 2012 and 2022, peaking in 2020–2021, with 62 more publications than the previous decade. The University of Chinese Academy of Sciences (UCAS), Northwest A&F University (NWAFU), and Chinese Academy of Agricultural Sciences (CAAS) are the top three research institutions in this field. Researchers are mainly located in China, The United States of America (USA), and India. Furthermore, the research in this field is primarily concerned with the screening of functional microorganisms, the determination of functional enzymes, and the analysis of microplastic biodegradation processes and mechanisms. These studies have revealed that the existing functional microorganisms for microplastic biodegradation are bacteria, predominantly Proteobacteria and Firmicutes; fungi, mainly Ascomycota; and some intestinal microorganisms. The main enzymes secreted in the process are hydrolase, oxidative, and depolymerization enzymes. Microorganisms degrade microplastics through the processes of colonization, biofilm retention, and bioenzymatic degradation. These studies have elucidated the current status of and problems in the microbial degradation of microplastics, and provide a direction for further research on the degradation process and molecular mechanism of functional microorganisms.

Keywords: microplastic; biodegradation; bibliometric; mechanism research

1. Introduction

"Microplastics" is used to describe plastic fibers, particles, and films that are less than 5 mm in size [1,2]. They are found in the air [3,4], oceans [5–9], soils [10,11], sediments, and surface waters [12–17] around the world, and have even been found in the sparsely populated Antarctic [18] and Arctic [19]. Some studies reported that the concentration of microplastics in surface water is between 10^{-5} and 10^5 sheets/m^3 and up to 67,500 mg/kg in soil [20–24]. Although there has been a great deal of research into their detection in the environment, there are difficulties in quantifying microplastics and nanoplastics, such as microplastics being too small to be fully captured and a lack of consistent standards for detection in different environments. In addition, a large number of microplastics in the environment are prone to migrate and transform in different environmental media due to their small particle size and low density, and release harmful substances such as phthalates (PAEs), polybrominated diphenyl ethers (PBDEs), and biphenyl A (BPA) in the process of migration and transformation [25]. Microplastics may also be compounded with persistent organic pollutants (POPs) and antibiotics to form composite pollutants in the ecosystem, thus causing harm to plants, animals, and microorganisms, and migrating into humans through the nutrient chain [26].

Due to the widespread presence of microplastics and their hazards, their removal from the environment has received considerable attention in recent years. The main removal methods include physical, chemical, and biological methods. The term "biological methods" refers to the use of microorganisms present in nature to remove microplastics from the environment. This is achieved through their metabolic activities, which further break up and split microplastics, decompose them into CO_2 and H_2O, and achieve inorganic mineralization. Biological degradation has been shown to be more effective than other methods, improving safety and environmental protection and contributing to the self-purification capacity of the contaminated environment. In terms of microplastic types, biodegradation focuses on both biodegradable and non-biodegradable microplastics. Biodegradable microplastics are environmentally friendly and can be completely degraded by microorganisms, while non-biodegradable microplastics degrade at a slower rate, with large differences in degradation efficiency and sometimes incomplete degradation. Common biodegradable microplastics are polylactic acid (PLA) and polybutylene terephthalate (PBAT). They are chemically bound to react readily in soil, water, and other environments, with microorganisms producing monomers through metabolic reactions. Common microorganisms include *Bacillaceae*, *Micromonosporaceae*, *Pseudonocardiaceae*, *Streptosporangiaceae*, and *Thermoactinomycetaceae*. In addition, under industrial composting conditions at 40 degrees Celsius, 80% of modified PLA fiber microplastics are completely degradable within a week, and Polycaprolactone (PCL) can be degraded in two days [27]. Recent studies have demonstrated that the metabolic processes of bacteria, fungi, and intestinal microorganisms facilitate the degradation of microplastics, rendering them a potential solution and effective mitigation measure for microplastic pollution [20,28]. *Exiguobacterium* sp. YT2 isolated from Tenebrio molitor larvae by Yang et al. formed biofilm on polystyrene (PS) after 28 days and reduced the weight by 7.4% in 60 days [29]. Similarly, Devi et al. isolated *Bacillus parapsilosis* (BP) and *Bacillus cereus* (BC) from the Vaigai River, India, and reduced the weight of Polypropylene (PP) and polyethylene (PE) by 78.99% and 63%, respectively [30]. Paco et al. found that the marine fungus *Zalerion maritimum* could use, and thus reduce the size and mass of, PE. The removal rate was as high as 43% after a 14-day degradation process [31].

In summary, current research has identified a large number of functional microorganisms and enzymes and explored the changes that occur in microplastics throughout time. However, there is still a lack of insight into the molecular mechanisms of biodegradation, the location of degradation, and the risk of degradation products. There is a need to comprehensively summarize current research, further explore the current knowledge on biodegradation by comparing the degradation of microplastics by different microorganisms, and study and understand the potential mechanisms of microplastic biodegradation in ecosystems in order to develop effective methods of microplastic removal and lay the foundation for future research.

Therefore, this review presents the first comprehensive analysis of microplastic biodegradation and its mechanisms using a bibliometric approach. Current research on microplastic biodegradation and its mechanisms was extracted from the published literature. Based on the number of studies available, their co-citation frequency, and keywords, the research hotspots of microplastic biodegradation and future research trends based on keyword clustering were derived to provide references for microplastic biodegradation research.

2. Materials and Methods

2.1. Data Retrieval

With "biodegradation of microplastics", "microplastic biodegradation process", and "microplastic biodegradation mechanism" as the keywords, four databases were comprehensively searched from inception to 2022: American Chemical Society (ACS), Elsevier, Springer Link, and American Society for Microbiology (ASM). No restrictions were implemented on the language, document type, data category, or year of publication. The data obtained from the four databases were consolidated, and a preliminary selection of identified publications was then undertaken. Since this study focuses on microplastics that are formed from large plastic particles through weathering and abrasion and exist in the environment for a long period of time, it excluded records of biodegradable microplastics produced by plastic products whose biodegradability was taken into account during the manufacturing process and studies focusing on microplastic–microorganism interactions and their impact on the ecosystem. For the articles that could not be identified, they were further filtered by reading the content.

2.2. Analysis Method

By means of bibliometric measures, we measured and visually analyzed the identified papers in the field of microplastic biodegradation, including the number of citations, top authors, active countries, institutions, and keywords. The articles were counted in Excel and categorized and mapped by type. A co-occurrence analysis of issuing countries, institutions, authors, and keywords was performed in CiteSpace (version 5.6. R 3) to count the number of articles published by different countries, institutions, and authors; to show the connection between them; and to show the hotspots of research in recent years in the field through co-occurring keywords and emergent. We used a table in VOSviewer (version 1.6.15) to count the co-cited journals and visualize the co-cited references to highlight the key contributions in the field.

3. Results

3.1. Paper Type and Quantity

A total of 545 papers on microplastic biodegradation and its mechanisms were identified and examined, and 331 were retained after excluding irrelevant and redundant records. First, the statistics of the included studies were ranked and analyzed to reach preliminary conclusions. There were six types of publications in the search results (Figure 1a). Most of these were research articles (73.19%), followed by review papers (22.83%). Research articles and comments may reflect the most recent trends and changes in research on the biological governance and mechanisms of microplastics.

By examining the relationship between the timing and the number of publications, it is possible to reveal the past status of the research field and predict its future direction [32]. This strategy was used to improve the understanding of microplastic biodegradation and its mechanisms. As shown in Figure 1b, the research phase can be divided into two based on the characteristics of the variation in the annual volume of studies on microplastics: in the first stage, the increase in publications with time from 2012 to 2018 was stable, and the annual increase was low for the slow development stage; in the second stage, the number of publications increased rapidly from 2018 to 2022. By 2021, the number of global research articles on the biological governance and mechanisms of microplastics exceeded 100 for the first time, reaching 127, doubling the number of publications from the previous year. This may be related to the fact that microplastics were identified as a new pollutant in 2019 and their presence has been found in the human body [33].

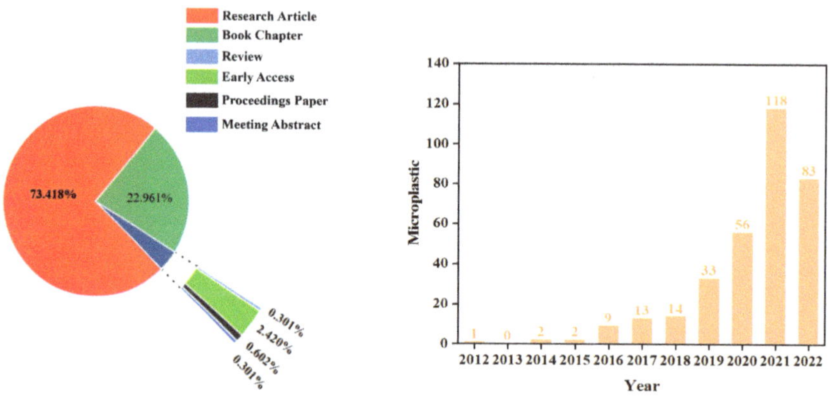

(a) (b)

Figure 1. Paper type and quantity: (**a**) percentages of the types of the research records on microplastic biological management and mechanisms of microplastics; (**b**) number of studies published from 2012 to 2022 on the biological management and mechanisms of microplastics.

3.2. Authors, Countries, and Institutions

To some extent, the number of publications published in specific research areas can reflect the impact on their field [34]. The top ten authors and co-cited authors were studied (Table 1). Currently, most authors in the field of microplastic biodegradation and mechanism research are from China, among whom Yong Zhang has the most papers.

Table 1. The top 10 authors and co-cited authors of microplastics research (N%).

Rank	Author	Country	N (%)	Co-Cited Author	Country	Citations
1	Zhang Y	China	17 (3.27%)	Lin Y	China	356
2	Li Y	China	12 (2.31%)	Li C	China	264
3	Wang J	China	12 (2.31%)	Li X	China	212
4	Chen J	China	9 (1.73%)	Li C	China	157
5	Chen Y	China	8 (1.54%)	Li X	China	119
6	Sun Y	China	8 (1.54%)	Li X	China	112
7	Zhang S	China	8 (1.54%)	Zhang S	China	76
8	Zhang Z	China	8 (1.54%)	Xie Y	China	72
9	Liu Y	China	7 (1.35%)	Lin Y	China	172
10	Wang X	China	7 (1.35%)	Li C	China	85

According to the analysis conducted using Citespace software, 81 countries and regions have published research papers on the biological governance and mechanisms of microplastics, among which the countries that have published within the top 10 countries are China, the United States of America, India, Italy, Germany, South Korea, England, Spain, Australia, and Greece. The top twenty countries in terms of publication number were studied, as were the links to the articles from each country (Figure 2a). The size of nodes symbolizes the number of papers; therefore, nodes become larger as the number of publications increases [35]. As can be seen from Figure 2a, the collaboration among countries is very close; China has the largest circle, representing the highest number of publications in the world. China's close collaborating countries are mainly the United States of America, the United Kingdom, Germany, and Australia.

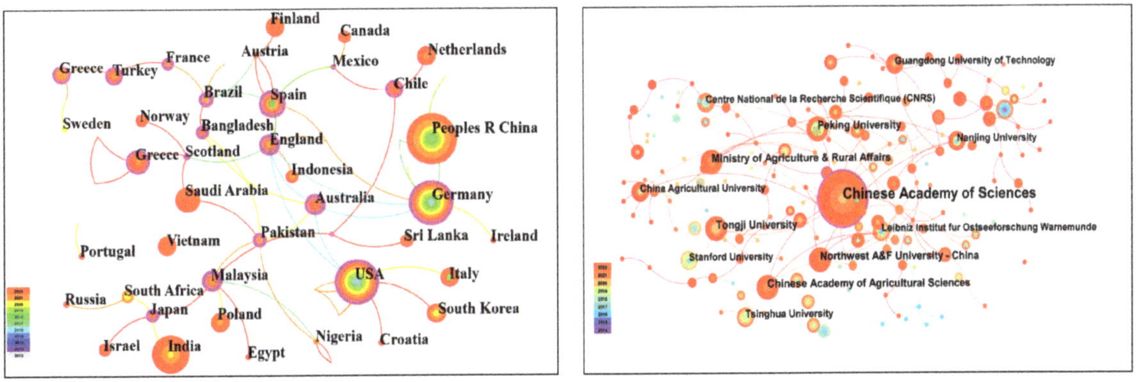

Figure 2. Country and institution network maps: (**a**) the top 20 productive countries; (**b**) the institutions conducting research on microplastic biological management and mechanisms of microplastics.

At present, 308 institutions have studied the biodegradation and mechanisms of microplastics. The Chinese Academy of Sciences represented the highest proportion at 8%, publishing 120 related papers, which accounted for 8%. The Chinese Academy of Agricultural Sciences came in second (5.8%) and Northwest A&F University came third (3.3%). Close cooperation was established between the institutions in each country. Among them, Nanjing University, the Chinese Academy of Sciences, and the University of the Chinese Academy of Sciences are connected with thick lines, indicating a close relationship (Figure 2b).

3.3. Co-Cited Journals and References

As shown in Table 2, Environmental Science & Technology is at the top of the list with 280 citations. Science of the Total Environment magazine ranks second and was cited 267 times. Of the five journals, the most cited number was concentrated in 2012.

Table 2. Top five co-cited journals.

Rank	Cited Journals	Count	Year
1	Environmental Science & Technology	280	2012
2	Science of the Total Environment	267	2016
3	Environmental Pollution	263	2012
4	Marine Pollution Bulletin	241	2012
5	Chemosphere	234	2012

As shown in Figure 3a, Geyer et al. (2017) published a paper on microplastics in the marine environment with the most citations at 87. There were five papers with more than six citations, published by Rummel et al. (2017), Auta et al. (2018), Paco et al. (2017), and Yuan et al. (2020). Compared with the citation frequency of microplastic-related papers published in the same period, the citation frequency of microplastic biodegradation is slightly lower, which may be related to the late start of research in this area. However, the citation frequency of each article on microplastic biodegradation was the highest at 48.17, indicating that the academic value of each article is high and the results are gradually receiving more attention (Figure 3b).

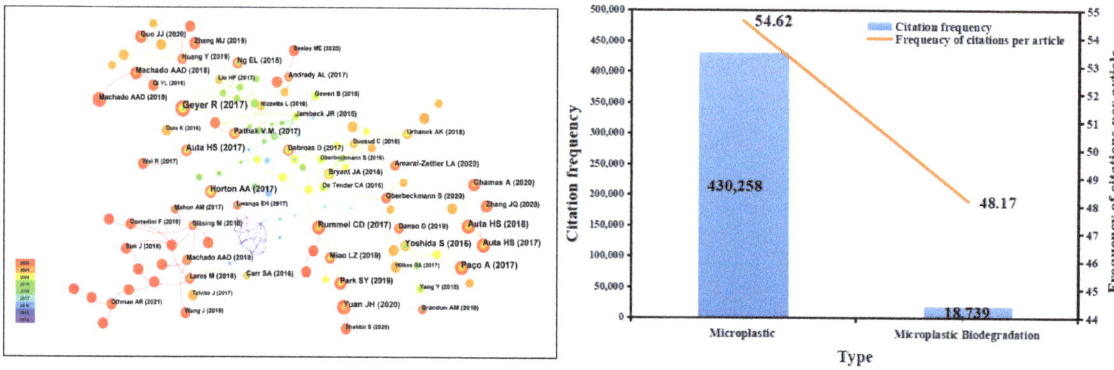

(a) (b)

Figure 3. Co-reference literature analysis: (**a**) analysis of microplastic co-references; (**b**) frequency of the cited literature.

3.4. Co-Occurrence and Burst Keywords Discussion

Keywords are words that have substantial meaning in expressing the central content of a paper and highlighting the key research directions in a field [36]. The keywords from 331 microplastic studies were summarized and counted. A density visualization of the chromatography based on keywords and hotspot intensity is displayed in Figure 4a, with warm and cold colors representing hot and cold spots, respectively. Microplastics, biodegradation, degradation, polyethylene, the microbial community, and microorganisms are common keywords with the highest density.

The color of the annual cycle reflects the periods before and after the appearance of each keyword. As the color changes from purple to red, the period in which each keyword appeared becomes more recent, and vice versa. By analyzing the key research directions from 2012 to 2022 (Figure 4b), it is identified that microplastic biodegradation and mechanism research has received extensive attention in the past five years, especially after 2019. This indicates the urgency of the microplastic pollution problem that needs to be solved and the new possibilities offered to this field with the development of novel biotechnologies.

The network diagram of research on the biological governance and mechanisms of microplastics for 2012–2022 (Figure 4c) illustrates the multiple topics involved in this field. There are nine clusters of keywords, in order from 0 to 8; the smaller the number, the more keywords there are. Each cluster is a composition of multiple closely related words, each with a different color. The main studies have focused on microplastics (7.87, 0.01); biodegradation (6.59, 0.05); marine pollution (5.15, 0.05); ecosystem services (5.15, 0.05); and fragmentation (5.15, 0.05).

Figure 4d facilitates the identification of the hot topics. The red lines in the diagram are heat bars, representing the time period of the strongest citation bursts [37]. The first keyword related to the marine environment was found in 2013. Biodegradation was a hot topic from 2013 to 2018, and has been the focus of research since 2017. Based on the above studies, we show that there are many studies on the microbial degradation of microplastics in the marine environment compared with the atmospheric and soil environments. The current degradation of bacteria and the studied degradation mechanisms are all found in the marine environment. However, the microplastic content in the atmospheric and soil environments cannot be underestimated, and understanding its biodegradation process and mechanism is very essential.

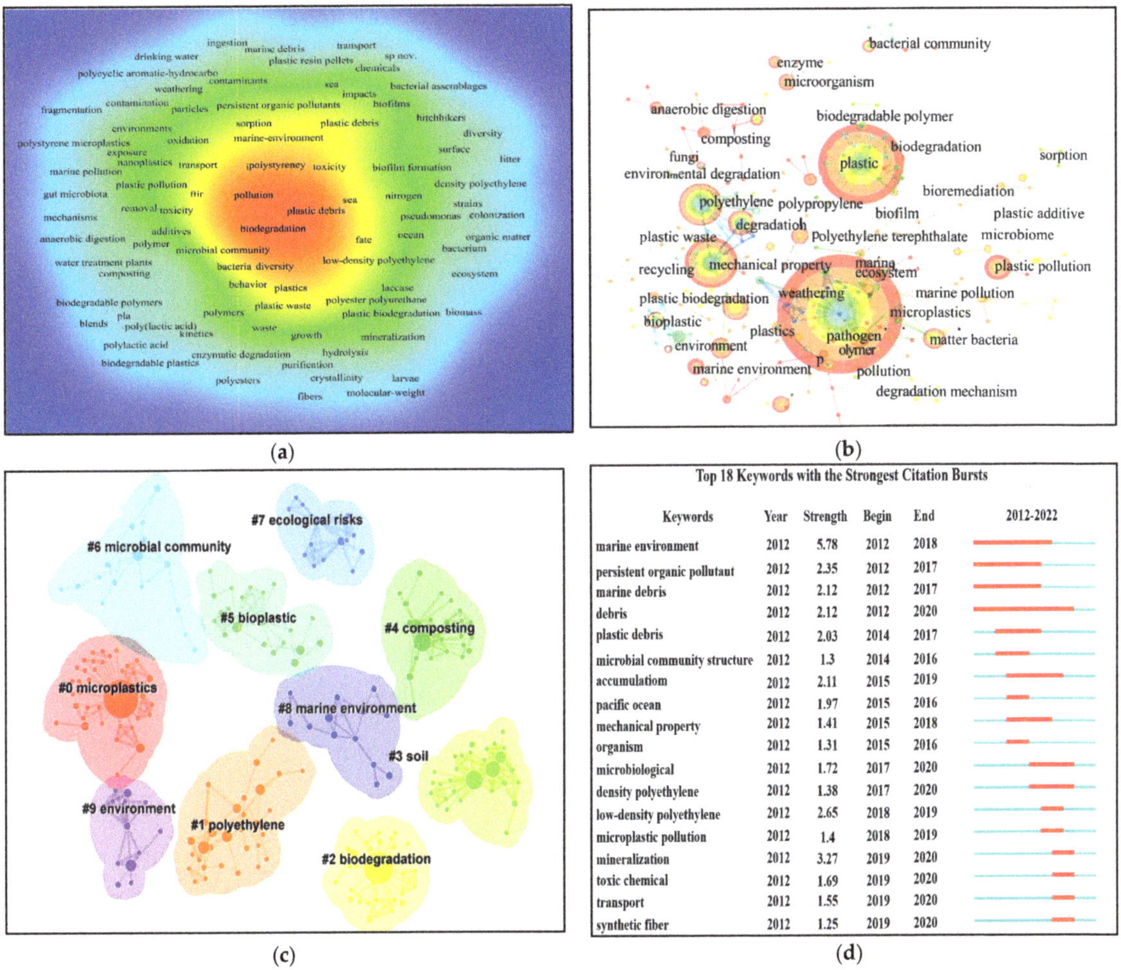

Figure 4. Keyword analysis: (**a**) visual map of the common occurrence density of keywords in the research on biological governance and mechanisms of microplastics from 2012–2022, the color of the annual cycle reflects the periods before and after the appearance of each keyword, as the color changes from purple to red, the period in which each keyword appeared becomes more recent, and vice versa; (**b**) network visualization map of co-occurrence keywords from 2012–2022, the color of the annual cycle reflects the periods before and after the appearance of each keyword, as the color changes from purple to red, the period in which each keyword appeared becomes more recent, and vice versa; (**c**) VOSvileviewer network diagram of research on biological governance and mechanisms of microplastics from 2012 to 2022, each node denotes a keyword and the keywords are clustered with different colours; (**d**) keywords with the strongest citation bursts, the red bars represent frequently cited keywords, while the green bars were infrequently cited keywords.

4. Discussion

4.1. The Status of Microplastic Biodegradation Research

A total of 331 articles were published from 2012 to 2022. Since 2012, the number of papers published has shown a trend of continuous growth. This indicates that the study of the biological governance and mechanisms of microplastics has developed rapidly during this period. The number of papers published in the past two years (2021–2022) is 62 more than in the previous decade. This may be due to the widespread concern about the dangers posed by microplastics. Microplastics have been shown to exist in plants and animals, entering through the human body along with the food chain, and in the human body through the blood circulation into multiple organs. It has been shown that when white blood cells in the human body swallow a certain amount of microplastics, they will die, releasing enzymes and causing inflammation. This then attacks more white blood cells, triggering continuous activation and leading to a variety of inflammatory and respiratory illnesses [38]. Therefore, a series of policies on microplastic control have been issued by various countries, such as the 2019 United Nations (UN) report, "Single-Use Plastics and Microplastic Legal Restrictions", "The Toxic Substances Control Act" of 2020, and "The New Pollutants Management Action Program" enacted by China in 2021, which provide direction and support for microplastic pollution management.

The top three countries in which articles were published were China, the United States of America, and India. Estonia, the Maldives, Costa Rica, and other countries had only one article. This indicates that China is the largest contributor to microplastic biodegradation research among developing countries.

In terms of the authors' countries, the top 10 authors are all Chinese. This shows that microplastic pollution is a hot topic of a lot of Chinese researchers in the field. However, we found that 112 authors (60.86% of the total authors) had published only one paper. This indicates that microplastic biodegradation is an emerging field, and although there are many researchers, there are still few results are still few and there is a lot of room for development.

The United States of America was the country that worked the closest with China. Kenya, Montenegro, Uganda, and other countries rarely collaborate with foreign countries. The process and mechanisms of microplastic biodegradation are a comprehensive field covering disciplines such as macrogenomics, enzymology, and chromatographic analysis. It is necessary to strengthen domestic and international cooperation in order to achieve further development in this field.

4.2. The Current Status of Functional Microorganism Development for Microplastic Biodegradation

Microplastics can provide a habitat for many special microbial communities, such as microplastic-degrading bacteria, that can change the ecological function of the ecosystem and play a role in microplastic degradation [39]. Microplastics are polymeric substances that have existed in the environment for a long time and do not degrade easily [40], remaining in the environment for a decade or longer. It has been found that some microorganisms in the ecosystem have significant degradation effects on microplastics. This includes bacteria and fungi, which can participate in the degradation of microplastics [28]. However, it is difficult for a single strain to completely degrade microplastics, and complexes of several types of bacteria are needed instead to be effective for the degradation of microplastics [20]. In addition to being widely distributed on microplastic surfaces, these functional microorganisms are also present in the digestive tract of some animals [41]. Whether it occurs through bacteria, fungi, or intestinal microorganisms, the main principle of microplastic biodegradation is a complete decomposition of microplastics into harmless inorganic substances through microbial metabolism. Microplastic biodegradation generally consists of three steps: First, microorganisms colonize the surface of the microplastic and form small particles. Second, cleavage occurs under the action of enzymes to form small molecular compounds. Third, the microplastics undergo mineralization and decomposition into CO_2 and H_2O. Bacteria have faster growth and reproduction, strong genetic manipulation, and clear metabolic pathways, which make it easier to elucidate the degradation process and mechanism in studies. Fungi have a higher degradation efficiency and strategy, their enzyme system is powerful, and their filamentous structure facilitates the secretion of extracellular enzymes, which can act on microplastics, enhance adsorption capacity, and generate biosurfactants; these characteristics are lacking in bacteria (Figure 5) [42].

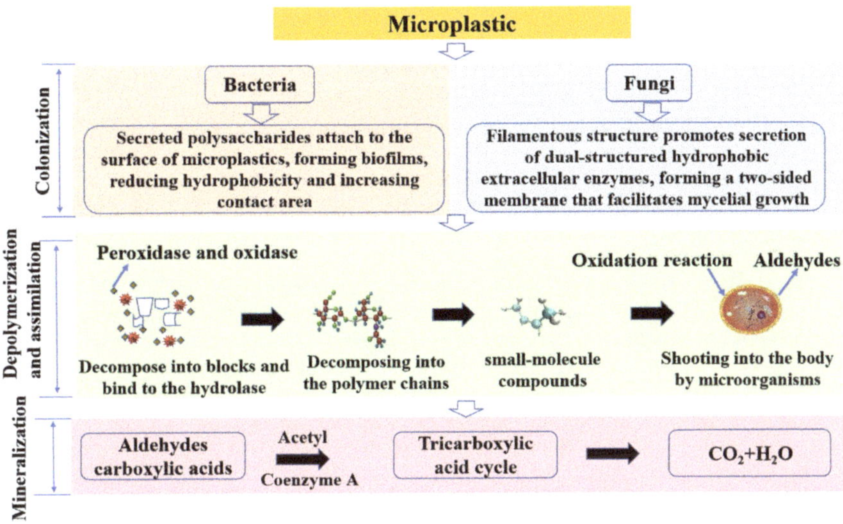

Figure 5. Microplastic biodegradation processes.

With the development of biotechnology, high-throughput sequencing technology has been widely applied to the study of microbial functional genes, which can be used to precisely probe the molecular mechanisms underlying certain functions. Firstly, it uses whole genome sequencing to obtain all the genetic information about degrading bacteria, and then the data are annotated and analyzed to see if the strains have gene-encoding enzymes with microplastic degradation potential and metabolic genes of microplastic metabolism-related pathways. Moreover, through genetic engineering techniques such as gene mutation, the degradation enzyme genes can be targeted for mutation to improve the degradation activity of the enzyme, and enzyme preparations can be obtained in bulk through mass replication in the strains. Finally, the degradation mechanisms of microplastics were investigated using genomics, secreted proteomics, and metabolomics.

4.2.1. Identified Microplastic Degradation of Bacteria and Degradation Mechanisms

Bacteria are one of the main groups of organisms and have the largest number of species of all organisms. There have been many reports on the bacterial degradation of microplastics. Plate separation is primarily used to isolate and purify degradative bacteria from samples in order to obtain single strains capable of degradation. Table 3A summarizes the degraded bacterial taxa reported to date. The selected microplastic-degrading bacteria belonged to four phyla, among which the families γ-Proteobacteria and Actinobacteria were the largest.

As more bacteria are found to reduce microplastic content, their degradation properties and effects on microplastics are receiving increased attention. Firstly, microplastics are eroded under the action of enzymes, oxidizing and weakening the structure, introducing polar groups into the hydrocarbon skeleton, and promoting a stronger attachment of microorganisms. The formation of small aliphatic hydrocarbons increases the possibility of bacterial uptake, and through several steps of dehydrogenation and C-C bond breaking, bacteria finally enter the tricarboxylic acid cycle, which provides the required carbon source and energy for bacterial metabolism.

The bacterial degradation of microplastics is a popular method. With the advancement of research, it has been found that the microplastic weight loss rate is low when bacteria are used for degradation, approximately 0–15%. In addition, the use of bacteria to degrade microplastics is a relatively long process, approximately 0–3 months. Therefore, shortening the degradation process of microplastics, improving the degradation rate of bacterial microorganisms, optimizing culture conditions, and developing improved degradation strains will be the focus of research in the future.

4.2.2. Identified Microplastic Degradation and Degradation Mechanisms in Fungi

Fungi are a large group, and their filamentous structure promotes the secretion of extracellular enzymes that can enable them to be better adapted to different environments for polymer degradation, such as dissolved organics [43,44]. Fungi seem to degrade polyethylene at a higher rate compared to bacteria [45]. As shown in Table 3A, the fungi that can degrade microplastics mainly belong to Ascomycota and Aspergillus, including *Aspergillus*, *Penicillium*, and *Rhizopus*.

Microplastic degradation by fungi involves the secretion of extracellular enzymes in the presence of filamentous structures. The dual structure of the hydrophobic protein produced by fungi acts on the surface of microplastics to form an amphipathic film that promotes the attachment of mycelium [46], which decreases the structural properties of microplastics [47] and depolymerizes to form oligomers and monomers [48]. Through a membrane transport system, these small molecules pass through a semipermeable membrane and are assimilated by the fungus [49].

Fungi have a high degradation efficiency, and their use in microplastic degradation is becoming a new research trend. The next step is to focus more efforts on screening and studying functional fungi that can degrade refractory microplastics. This will provide theoretical and technical support for the biological governance of the environment, aiming to reduce the impact of microplastics [28,50].

4.2.3. Microplastic Degradation by Insect Gut Microorganisms and Its Mechanism

Some insects can chew and eat beeswax or plastic products and use them as their sole carbon source, providing a powerful biological resource for the biodegradation of microplastics [51]. Eight insect species have been reported to feed on and degrade plastic, and most studies have focused on yellow mealworms, barley worms, and wax borer larvae (Table 3B). Larval-stage yellow mealworms, earthworms, and wax borers swallow and, with the help of the microorganisms in their bodies, degrade various plastic polymers [52]. The microorganisms in these insects' bodies enable them to degrade microplastics. Therefore, in addition to isolating biodegradable microplastics directly from plastic rings, screening for microplastic-degrading bacteria in these insects has become a viable option.

4.2.4. Microplastic Degradation Enzymes

Fungi and bacteria are able to degrade microplastics through the secretion of various degrading enzymes. Many scholars have begun to degrade microplastics through direct enzyme synthesis. Microplastic-degradative enzymes can be classified as extracellular enzymes and intracellular enzymes according to their position. Microplastic enzymes involved in long-carbon-chain separation, which make plastic polymers form oligomers or dimers, can be absorbed into the cell. However, since research on microplastic-degrading enzymes is fairly new, few studies have investigated the enzymes during the microbial degradation process, and the focus has mainly been on extracellular enzymes. Currently, the enzymes found to play important roles in the degradation process of microplastics are proteases, lipase, keratinase, laccase, manganese peroxidase, lignin peroxidase, and alkane hydroxylase [53,54]. Polymers can undergo enzymatic biodegradation. However, for the enzymatic degradation of microplastics, five important factors must be addressed: surface topography, crystallinity, water absorption, polymer chain orientation, and reaction temperature. Studies have been reported on the application of polyethylene terephthalate hydrolases in fibers, films, and nanoparticles. The enzymatic hydrolysis of other microplastics has also been the target of previous studies that focused on the enzymatic surface modification of polymer fibers (changing the functional groups on the surface of microplastic fibers) and the treatment of microplastic waste (degradation of fiber building blocks). In other words, surface modification improves the wettability, fastness, dyeing, and pilling resistance of microplastics and ultimately the hydrophilicity of plastics. The surface area-to-volume ratio plays an important role in the degradation process; the higher or lower this ratio is, the more easily the polymer can degrade compared to fiber or film [55–57].

4.3. Current Status of Development of Biodegradation of Different Types of Microplastics

Polyethylene (PE), Polystyrene (PS), and Polyethylene Terephthalate (PET) are common plastic materials that are widely used in daily life. Microplastics formed after crushing are degraded under the action of microorganisms, and the degradation processes and mechanisms of different microplastics differ significantly due to their different compositions. In addition, the degradation process and mechanism will also change under the influence of environmental factors in actual degradation. PE microplastics are straight-chain alkanes with a high degree of polymerization and strong flexibility. Their degradation process consists of two stages: the first is the depolymerization process under the action of peroxidase, which depolymerizes the polymer into oligomers with 10–50 C atoms; the second is the mineralization stage after hydroxylation, which achieves the degradation of PE microplastics through the tricarboxylic acid cycle. PS microplastics are aromatic polymers that degrade under the action of a series of enzyme-catalyzed reactions such as dehydrogenation, deoxygenation, and hydroxylation. Microbial degradation is accomplished through two different metabolic pathways, namely, the styrene-ethyleneglycol-3-ethylene diphenol pathway and the styrene oxides-phenylacetaldehyde pathway. The biodegradation of PET microplastics starts with the PETase enzyme, which generates ethylene glycol (EG), monohydroxyethyl tetrahydronaphthalene (MHET), and Bis(2-hydroxyethyl) terephthalate (BHET), and then moves to degradation by terephthalic acid (TPA), ethylene glycol (EG), and ultimately the tricarboxylic acid cycle (TCA) through oxidase, which completes the biodegradation of PET microplastics [58].

Table 3. (**A**) Typical bacteria and fungi with microplastic degradation functions. (**B**) Microplastic-degrading insects and their internal microorganisms involved in degradation.

		(A)			
Taxon	**Isolate**	**Time**	**% of Degradation**	**Microplastic**	**Reference**
Bacteria-Proteobacteria	*Acinetobacter pitti IRN19*	4 w	26.8 ± 3.04	LDPE	[59,60]
	Pseudomonas	8 w	5.95 ± 0.03	PE	
	Alcanivorax borkumensis	80 d	3.5 ± 0.34	LDPE	
	Pseudomona saeruginosa	60 d	51	LDPE	
	Achromobacter	150 d	9	HDPE	
	Comamonas	90 d	46.7	PE	
	Enterobacter	30 d	12	HDPE	
	Microbulbifer	90 d	36.88 ± 1.25	HDPE	
	Stenotrophomonas	90 d	18.18 ± 0.69	HDPE	
Bacteria-Firmicutes	Geobacillus stearothermophilus FAFU011	56 d	4.20	PS	
	Paenibacillus sp.	60 d	14.7	PE	
	Lysinibacillus	90 d	36.88 ± 1.25	HDPE	
Bacteria-Actinobacteria	*Rhodococcus* sp. *IR-SGS-T6*	60 d	1.58	LDPE	
	Nocardia sp. *IR-SGS-T3*	60 d	1.58	LDPE	
	Streptomyces sp.	60 d	1.58	LDPE	
	Kocuria	60 d	14.7	PE	
Bacteria-Bacteroidetes	*Sphingobacterium*	28 d	3.04	LDPE	
Fungi-Ascomycota	*Aspergillus* sp.	30 d	-	PP/PBAT	
	Fusarium spp.	28 d	30	LDPE	
	Cladosporium pseudocladosporioides T1.PL.1	14 d	65	PU	
	Penicillium sp.	30 d	-	PP/PBAT	
	Trichoderma	28 d	-	PE	
	Alternaria	120 d	52.02	PE	
	Lasiodiplodia	90 d	-	LDPE	

Table 3. *Cont.*

(B)			
Insect Name	**Microplastic**	**Degradation Effect**	**Function Microbial**
Waxworms, Indian Mealmoths	Polyethene	Chew and feed on polyethylene; microbial degradation in the body	*Nitrobacteria asburiae YT1, Bacillus* sp. *YPI*
Zophobas Morio	Polystyrene	Using polystyrene as the carbon source, the weight and number in feces are reduced, and microbial degradation in vivo	*Klebsiella, Citrobacter*
Mealworms	Polyethene, polystyrene	Using these two microplastics as carbon sources, there is microbial degradation in the intestine with a combined effect	*Citrobacter* sp., *Kosakonia* sp.
Dark Mealworms	Polystyrene	Using polystyrene as a carbon source, polystyrene depolymerization was found in the gut	*Enterobacteriaceae*
Waxworms	Polyethene	Microbial degradation bacteria were identified in the gut	*Enterobacter* sp. *DI*
Yellow Mealworms	Polystyrene	Microbial degradation bacteria were identified in the gut and feces	*Aspergillus niger KHJ-1*
Earthworm	Low-density polyethylene	Microbial degradation bacteria were identified in the gut	*Rhodococcus jostii, Mycobacterium*

5. Conclusions

This review employs bibliometric analysis to select articles on microplastic biodegradation and mechanisms from 2012 to 2022 and systematically summarizes the current research hotspots and progress in microplastic biodegradation and its mechanisms through the study of annual publication volume, countries, institutions, authors, and keywords. Based on the results, 2018 to 2022 was the rapid growth phase of publications on microplastic biodegradation. In this period, 81 countries and 308 institutions started research in this field, with significant contributions from Chinese scholars and research institutions. According to the statistics, China's research volume accounted for 32.7%, the highest percentage. Secondly, the top three authors and institutions are from China, of which the Chinese Academy of Sciences represents 8% of the publications, works closely with other institutions at in China and abroad, and is the most active in academic exchanges. Zhang Yong is the author of the largest volume of publications and Li X is an author with whom many authors have cooperated closely. However, at present, most of these authors are concentrated in China, and communication with foreign scholars needs to be strengthened. From the keyword network diagram, microplastics, biodegradation, and the marine environment have been research hotspots in recent years. Although we have carried out a great deal of related research, problems are still encountered in biodegradation, such as unknown processes and unclear mechanisms, which need to be solved urgently. Based on the current research progress, the following aspects should be focused on in the future:

1. Only a few types of degrading strains have been identified. Fungi are currently only found to degrade polyvinyl chloride (PVC). The role of bacteria and enzymes needs to be investigated further.
2. Holistic knowledge of the degradation process needs to be obtained. Current research on the biodegradation and mechanisms of microplastics mainly focuses on the marine environment, with less data available for the soil, water, and atmospheric environments. This limits a holistic understanding of microplastic degradation and makes describing its mechanisms more challenging. Therefore, there is a need to investigate the processes by which microplastic degradation occurs in different environments and the links between them.

Author Contributions: Investigation and writing—original draft; Y.C. and J.B.; supervision, conceptualization, writing—review and editing: Y.H. and J.L.; writing—original draft: Y.M., W.F., Y.D. and Y.Y. All authors have read and agreed to the published version of the manuscript.

Funding: This work was supported by the Project of Science and Technology of Inner Mongolia Province, Inner Mongolia Autonomous Region, China (2022YFHH0044), the Inner Mongolia Foundation, Inner Mongolia Autonomous Region, China (2023LHMS05027), the Basic Research Operating Expenses Program for Colleges and Universities directly under the Autonomous Region, Inner Mongolia Autonomous Region, China (JY20220042), and the Innovation and Practice Training Program for Undergraduates of the Chinese Academy of Sciences from Chinese Academy of Sciences, Beijing, China (20234001093).

Institutional Review Board Statement: Not applicable.

Informed Consent Statement: Not applicable.

Data Availability Statement: Data will be made available on request.

Conflicts of Interest: The authors declare that they have no known competing financial interests or personal relationships that could have appeared to influence the work reported in this paper.

Abbreviations

ACS	American Chemical Society
ASM	American Society for Microbiology
UCAS	Chinese Academy of Sciences
NWAFU	Northwest A&F University
CAAS	Chinese Academy of Agricultural Sciences
USA	The United States of America
PAEs	Phthalates
PBDEs	Polybrominated Diphenyl Ethers
BPA	Biphenyl A
POPs	Persistent Organic Pollutants
PLA	Polylactic Acid
PBAT	Polybutylene Terephthalate
PCL	Polycaprolactone
PS	Polystyrene
BP	Bacillus parapsilosis
BC	Bacillus cereus
PP	Polypropylene
PE	Polyethylene
UN	United Nations
PET	Polyethylene Terephthalate
EG	Ethylene Glycol
MHET	Monohydroxyethyl Tetrahydronaphthalene
BHET	Bis(2-hydroxyethyl) terephthalate
TPA	Terephthalic Acid
TCA	Ultimately The Tricarboxylic Acid Cycle
LDPE	Low Density Polyethylene
HDPE	High Density Polyethylene
PU	Polyurethane

References

1. Peng, G.; Xu, P.; Zhu, B.; Bai, M.; Li, D. Microplastics in freshwater river sediments in Shanghai, China: A case study of risk assessment in mega-cities. *Environ. Pollut.* **2018**, *234*, 448–456. [CrossRef]
2. Zhang, G.-X.; Gu, J.-D. Biodegradability of plastics: The pitfalls. *Appl. Environ. Microbiol.* **2017**, *2*, 59–61. [CrossRef]
3. Dris, R.; Gasperi, J.; Mirande, C.; Mandin, C.; Guerrouache, M.; Langlois, V.; Tassin, B. A first overview of textile fibers, including microplastics, in indoor and outdoor environments. *Environ. Pollut.* **2017**, *221*, 453–458. [CrossRef]
4. Dris, R.; Gasperi, J.; Saad, M.; Mirande, C.; Tassin, B. Synthetic fibers in atmospheric fallout: A source of microplastics in the environment? *Mar. Pollut. Bull.* **2016**, *104*, 290–293. [CrossRef]
5. Fendall, L.S.; Sewell, M.A. Contributing to marine pollution by washing your face: Microplastics in facial cleansers. *Mar. Pollut. Bull.* **2009**, *58*, 1225–1228. [CrossRef]
6. Cole, M.; Lindeque, P.; Halsband, C.; Galloway, T.S. Microplastics as contaminants in the marine environment: A review. *Mar. Pollut. Bull.* **2011**, *62*, 2588–2597. [CrossRef] [PubMed]

7. Ng, K.; Obbard, J. Prevalence of microplastics in Singapore's coastal marine environment. *Mar. Pollut. Bull.* **2006**, *52*, 761–767. [CrossRef] [PubMed]
8. Wang, J.; Tan, Z.; Peng, J.; Qiu, Q.; Li, M. The behaviors of microplastics in the marine environment. *Mar. Environ. Res.* **2016**, *113*, 7–17. [CrossRef]
9. Pan, T.; Liao, H.; Yang, F.; Sun, F.; Guo, Y.; Yang, H.; Feng, D.; Zhou, X.; Wang, Q. Review of microplastics in lakes: Sources, distribution characteristics, and environmental effects. *Carbon Res.* **2023**, *2*, 25. [CrossRef]
10. Piehl, S.; Leibner, A.; Löder, M.G.J.; Dris, R.; Bogner, C.; Laforsch, C. Identification and quantification of macro- and microplastics on an agricultural farmland. *Sci. Rep.* **2018**, *8*, 17950. [CrossRef]
11. Guo, J.-J.; Huang, X.-P.; Xiang, L.; Wang, Y.-Z.; Li, Y.-W.; Li, H.; Cai, Q.-Y.; Mo, C.-H.; Wong, M.-H. Source, migration and toxicology of microplastics in soil. *Environ. Int.* **2020**, *137*, 105263. [CrossRef] [PubMed]
12. Srinivasa Reddy Reddy, M.; Basha, S.; Adimurthy, S.; Ramachandraiah, G. Description of the small plastics fragments in marine sediments along the Alang-Sosiya ship-breaking yard, India. *Estuar. Coast. Shelf Sci.* **2006**, *68*, 656–660. [CrossRef]
13. Van Cauwenberghe, L.; Vanreusel, A.; Mees, J.; Janssen, C.R. Microplastic pollution in deep-sea sediments. *Environ. Pollut.* **2013**, *182*, 495–499. [CrossRef] [PubMed]
14. Cerón, M.R.; Izquierdo, M.; Alegret, N.; Valdez, J.A.; Rodríguez-Fortea, A.; Olmstead, M.M.; Balch, A.L.; Poblet, J.M.; Echegoyen, L. Reactivity differences of $Sc_3N@C_{2n}$ (2n = 68 and 80). Synthesis of the first methanofullerene derivatives of $Sc_3N@D_{5h}$-C_{80}. *Chem. Commun.* **2016**, *52*, 64–67. [CrossRef]
15. McDermid, K.J.; McMullen, T.L. Quantitative analysis of small-plastic debris on beaches in the Hawaiian archipelago. *Mar. Pollut. Bull.* **2004**, *48*, 790–794. [CrossRef] [PubMed]
16. Andrady, A.L. Microplastics in the marine environment. *Mar. Pollut. Bull.* **2011**, *62*, 1596–1605. [CrossRef] [PubMed]
17. Li, L.; Geng, S.; Wu, C.; Song, K.; Sun, F.; Visvanathan, C.; Xie, F.; Wang, Q. Microplastics contamination in different trophic state lakes along the middle and lower reaches of Yangtze River Basin. *Environ. Pollut.* **2019**, *254*, 112951. [CrossRef]
18. Barnes, D.K.; Walters, A.; Gonçalves, L. Macroplastics at sea around Antarctica. *Mar. Environ. Res.* **2010**, *70*, 250–252. [CrossRef] [PubMed]
19. Bergmann, M.; Sandhop, N.; Schewe, I.; D'hert, D. Observations of floating anthropogenic litter in the Barents Sea and Fram Strait, Arctic. *Polar Biol.* **2016**, *39*, 553–560. [CrossRef]
20. Yoshida, S.; Hiraga, K.; Takehana, T.; Taniguchi, I.; Yamaji, H.; Maeda, Y.; Toyohara, K.; Miyamoto, K.; Kimura, Y.; Oda, K. A bacterium that degrades and assimilates poly(ethylene terephthalate). *Science* **2016**, *351*, 1196–1199. [CrossRef]
21. Browne, M.A.; Crump, P.; Niven, S.J.; Teuten, E.; Tonkin, A.; Galloway, T.; Thompson, R. Accumulation of Microplastic on Shorelines Woldwide: Sources and Sinks. *Environ. Sci. Technol.* **2011**, *45*, 9175–9179. [CrossRef]
22. Liebezeit, G.; Dubaish, F. Microplastics in Beaches of the East Frisian Islands Spiekeroog and Kachelotplate. *Bull. Environ. Contam. Toxicol.* **2012**, *89*, 213–217. [CrossRef]
23. Nor, N.H.M.; Obbard, J.P. Microplastics in Singapore's coastal mangrove ecosystems. *Mar. Pollut. Bull.* **2014**, *79*, 278–283. [CrossRef]
24. Desforges, J.-P.W.; Galbraith, M.; Dangerfield, N.; Ross, P.S. Widespread distribution of microplastics in subsurface seawater in the NE Pacific Ocean. *Mar. Pollut. Bull.* **2014**, *79*, 94–99. [CrossRef]
25. Zhao, X.; Wu, X.; White, J.C.; Tang, Z.; Wu, F. Micro/nanoplastics contamination of the terrestrial environment: Exposure routes, dose, and co-contaminants complicate the risk calculus. *Carbon Res.* **2023**, *2*, 24. [CrossRef]
26. Nizzetto, L.; Futter, M.; Langaas, S. Are Agricultural Soils Dumps for Microplastics of Urban Origin? *Environ. Sci. Technol.* **2016**, *50*, 10777–10779. [CrossRef]
27. Qin, M.; Chen, C.; Song, B.; Shen, M.; Cao, W.; Yang, H.; Zeng, G.; Gong, J. A review of biodegradable plastics to biodegradable microplastics: Another ecological threat to soil environments? *J. Clean.* **2021**, *312*, 127816. [CrossRef]
28. Yuan, J.; Ma, J.; Sun, Y.; Zhou, T.; Zhao, Y.; Yu, F. Microbial degradation and other environmental aspects of microplastics/plastics. *Sci. Total Environ.* **2020**, *715*, 136968. [CrossRef]
29. Xiang, P.; Zhang, T.; Wu, Q.; Li, Q. Systematic Review of Degradation Processes for Microplastics: Progress and Prospects. *Sustainability* **2023**, *15*, 12698. [CrossRef]
30. Devi, K.N.; Raju, P.; Santhanam, P.; Kumar, S.D.; Krishnaveni, N.; Roopavathy, J.; Perumal, P. Biodegradation of low-density polyethylene and polypropylene by microbes isolated from Vaigai River, Madurai, India. *Arch. Microbiol.* **2021**, *203*, 6253–6265. [CrossRef] [PubMed]
31. Li, Q.; Luo, Y.; Sha, A.; Xiao, W.; Xiong, Z.; Chen, X.; He, J.; Peng, L.; Zou, L. Analysis of synonymous codon usage patterns in mitochondrial genomes of nine Amanita species. *Front. Microbiol.* **2023**, *8*, 1134228. [CrossRef] [PubMed]
32. Gao, W.; Chen, Y.; Liu, Y.; Guo, H.-C. Scientometric analysis of phosphorus research in eutrophic lakes. *Scientometrics* **2015**, *102*, 1951–1964. [CrossRef]
33. Liu, Z.; You, X. Recent progress of microplastic toxicity on human exposure base on in vitro and in vivo studies. *Sci. Total Environ.* **2023**, *903*, 166766. [CrossRef]
34. Borrelle, S.B.; Ringma, J.; Law, K.L.; Monnahan, C.C.; Lebreton, L.; McGivern, A.; Murphy, E.; Jambeck, J.; Leonard, G.H.; Hilleary, M.A.; et al. Predicted growth in plastic waste exceeds efforts to mitigate plastic pollution. *Science* **2020**, *369*, 1515–1518. [CrossRef]
35. Li, H.; Jiang, H.-D.; Yang, B.; Liao, H. An analysis of research hotspots and modeling techniques on carbon capture and storage. *Sci. Total Environ.* **2019**, *687*, 687–701. [CrossRef] [PubMed]

36. Qin, F.; Du, J.; Gao, J.; Liu, G.; Song, Y.; Yang, A.; Wang, H.; Ding, Y.; Wang, Q. Bibliometric Profile of Global Microplastics Research from 2004 to 2019. *Int. J. Environ. Res. Public Health* **2020**, *17*, 5639. [CrossRef] [PubMed]
37. Shi, J.; Shi, S.; Shi, S.; Jia, Q.; Yuan, G.; Chu, Y.; Wang, H.; Hu, Y.; Cui, H. Bibliometric analysis of potassium channel research. *Channels* **2020**, *14*, 18–27. [CrossRef] [PubMed]
38. Lin, Y.-D.; Huang, P.-H.; Chen, Y.-W.; Hsieh, C.-W.; Tain, Y.-L.; Lee, B.-H.; Hou, C.-Y.; Shih, M.-K. Sources, Degradation, Ingestion and Effects of Microplastics on Humans: A Review. *Toxics* **2023**, *11*, 747. [CrossRef]
39. Huang, Y.; Zhao, Y.; Wang, J.; Zhang, M.; Jia, W.; Qin, X. LDPE microplastic films alter microbial community composition and enzymatic activities in soil. *Environ. Pollut.* **2019**, *254*, 112983. [CrossRef]
40. Krueger, M.C.; Harms, H.; Schlosser, D. Prospects for microbiological solutions to environmental pollution with plastics. *Appl. Microbiol. Biotechnol.* **2015**, *99*, 8857–8874. [CrossRef]
41. Zhang, B.; Chen, L.; Chao, J.; Yang, X.; Wang, Q. Research Progress of Microplastics in Freshwater Sediments in China. *Environ. Sci. Pollut. Res.* **2020**, *27*, 31046–31060. [CrossRef] [PubMed]
42. Liang, R.; Sun, F.; Zhang, C.; Zhang, R.; Wang, H.; Wang, X. Progress of microplastic-microbial interactions in soil environment. *J. Bioeng.* **2023**, *39*, 500–515. [CrossRef]
43. Feng, W.; Wang, T.; Zhu, Y.; Sun, F.; Giesy, J.P.; Wu, F. Chemical composition, sources, and ecological effect of organic phosphorus in water ecosystems: A review. *Carbon Res.* **2023**, *2*, 12. [CrossRef]
44. Wang, T.; Feng, W.; Liu, J.; Fan, W.; Li, T.; Song, F.; Yang, F.; Liao, H.; Leppäranta, M. Eutrophication in cold-arid lakes: Molecular characteristics and transformation mechanism of DOM under microbial action at the ice-water interface. *Carbon Res.* **2024**, *3*, 42. [CrossRef]
45. Muhonja, C.N.; Makonde, H.; Magoma, G.; Imbuga, M. Biodegradability of polyethylene by bacteria and fungi from Dandora dumpsite Nairobi-Kenya. *PLoS ONE* **2018**, *13*, e0198446. [CrossRef] [PubMed]
46. Nauendorf, A.; Krause, S.; Bigalke, N.K.; Gorb, E.V.; Gorb, S.N.; Haeckel, M.; Wahl, M.; Treude, T. Microbial colonization and degradation of polyethylene and biodegradable plastic bags in temperate fine-grained organic-rich marine sediments. *Mar. Pollut. Bull.* **2016**, *103*, 168–178. [CrossRef] [PubMed]
47. Sánchez, C. Fungal potential for the degradation of petroleum-based polymers: An overview of macro- and microplastics biodegradation. *Biotechnol. Adv.* **2020**, *40*, 107501. [CrossRef] [PubMed]
48. Miri, S.; Saini, R.; Davoodi, S.M.; Pulicharla, R.; Brar, S.K.; Magdouli, S. Biodegradation of microplastics: Better late than never. *Chemosphere* **2022**, *286*, 131670. [CrossRef] [PubMed]
49. Ru, J.; Huo, Y.; Yang, Y. Microbial Degradation and Valorization of Plastic Wastes. *Front. Microbiol.* **2020**, *11*, 442. [CrossRef]
50. Zhang, A.; Hou, Y.; Wang, Q.; Wang, Y. Characteristics and polyethylene biodegradation function of a novel cold-adapted bacterial laccase from Antarctic sea ice psychrophile *Psychrobacter* sp. NJ228. *J. Hazard. Mater.* **2022**, *39*, 129656. [CrossRef]
51. Kannan, M.; Mubarakali, D.; Thiyonila, B.; Krishnan, M.; Padmanaban, B.; Shantkriti, S. Insect gut as a bioresource for potential enzymes—An unexploited area for industrial biotechnology. *Biocatal. Agric. Biotechnol.* **2019**, *18*, 101010. [CrossRef]
52. Amobonye, A.; Bhagwat, P.; Singh, S.; Pillai, S. Plastic biodegradation: Frontline microbes and their enzymes. *Sci. Total Environ.* **2021**, *759*, 143536. [CrossRef]
53. Dey, A.S.; Bose, H.; Mohapatra, B.; Sar, P. Biodegradation of Unpretreated Low-Density Polyethylene (LDPE) by *Stenotrophomonas* sp. and *Achromobacter* sp., Isolated from Waste Dumpsite and Drilling Fluid. *Front. Microbiol.* **2020**, *1*, 603210. [CrossRef]
54. Ahmed, T.; Shahid, M.; Azeem, F.; Rasul, I.; Shah, A.A.; Noman, M.; Hameed, A.; Manzoor, N.; Manzoor, I.; Muhammad, S. Biodegradation of plastics: Current scenario and future prospects for environmental safety. *Environ. Sci. Pollut. Res.* **2018**, *25*, 7287–7298. [CrossRef]
55. Kannahi, M.; Sudha, P. Screening of polythene and plastic degrading microbes from Muthupet mangrove soil. *J. Chem. Pharm. Res.* **2013**, *5*, 122–127.
56. Auta, H.S. Screening of Bacillus strains isolated from mangrove ecosystems in Peninsular Malaysia for microplastic degradation. *Environ. Pollut.* **2017**, *231*, 1552–1559. [CrossRef]
57. Lv, S.; Li, Y.; Zhao, S.; Shao, Z. Biodegradation of Typical Plastics: From Microbial Diversity to Metabolic Mechanisms. *Int. J. Mol. Sci.* **2024**, *25*, 593. [CrossRef]
58. Feng, W.; Deng, Y.; Cao, Y.; Liu, J.; Han, Y.; Liu, J.; Miao, Q.; Yang, F.; Zhu, Y.; Giesy, J.P. Biotechnology Remediation and Environmental Behavior of Microplastics in Soils: A Review. *Rev. Environ. Contam. Toxicol.* **2023**, *261*, 13. [CrossRef]
59. Dey, S.; Anand, U.; Kumar, V.; Kumar, S.; Ghorai, M.; Ghosh, A.; Kant, N.; Suresh, S.; Bhattacharya, S.; Bontempi, E.; et al. Microbial strategies for degradation of microplastics generated from COVID-19 healthcare waste. *Environ. Res.* **2023**, *216*, 114438. [CrossRef]
60. Gambarini, V.; Pantos, O.; Kingsbury, J.M.; Weaver, L.; Handley, K.M.; Lear, G. Phylogenetic distribution of plastic-degrading microorganisms. *Msystems* **2021**, *6*, e01112-20. [CrossRef]

MDPI AG
Grosspeteranlage 5
4052 Basel
Switzerland
Tel.: +41 61 683 77 34

Toxics Editorial Office
E-mail: toxics@mdpi.com
www.mdpi.com/journal/toxics

www.ingramcontent.com/pod-product-compliance
Lightning Source LLC
Chambersburg PA
CBHW062044220926
51850CB00010B/2652

* 9 7 8 3 7 2 5 8 7 5 6 3 4 *